Poisson Line Cox Process

Foundations and Applications to Vehicular Networks

Synthesis Lectures on Learning, Networks, and Algorithms

Editor
Lei Ying, *University of Michigan, Ann Arbor*

Editor Emeritus
R. Srikant, *University of Illinois at Urbana-Champaign*

Founding Editor Emeritus
Jean Walrand, *University of California, Berkeley*

Synthesis Lectures on Learning, Networks, and Algorithms is an ongoing series of 75- to 150-page publications on topics on the design, analysis, and management of complex networked systems using tools from control, communications, learning, optimization, and stochastic analysis. Each lecture is a self-contained presentation of one topic by a leading expert. The topics include learning, networks, and algorithms, and cover a broad spectrum of applications to networked systems including communication networks, data-center networks, social, and transportation networks. The series is designed to:

- Provide the best available presentations of important aspects of complex networked systems.

- Help engineers and advanced students keep up with recent developments in a rapidly evolving field of science and technology.

- Facilitate the development of courses in this field.

Poisson Line Cox Process: Foundations and Applications to Vehicular Networks
Harpreet S. Dhillon and Vishnu Vardhan Chetlur
2020

Age of Information: A New Metric for Information Freshness
Yin Sun, Igor Kadota, Rajat Talak, and Eytan Modiano
2019

Multi-Armed Bandits: Theory and Applications to Online Learning in Networks
Qing Zhao
2019

Network Games: Theory, Models, and Dynamics
Ishai Menache and Asuman Ozdaglar
2011

An Introduction to Models of Online Peer-to-Peer Social Networking
George Kesidis
2010

Stochastic Network Optimization with Application to Communication and Queueing Systems
Michael J. Neely
2010

Scheduling and Congestion Control for Wireless and Processing Networks
Libin Jiang and Jean Walrand
2010

Performance Modeling of Communication Networks with Markov Chains
Jeonghoon Mo
2010

Communication Networks: A Concise Introduction
Jean Walrand and Shyam Parekh
2010

Path Problems in Networks
John S. Baras and George Theodorakopoulos
2010

Performance Modeling, Loss Networks, and Statistical Multiplexing
Ravi R. Mazumdar
2009

Network Simulation
Richard M. Fujimoto, Kalyan S. Perumalla, and George F. Riley
2006

Poisson Line Cox Process: Foundations and Applications to Vehicular Networks
Harpreet S. Dhillon and Vishnu Vardhan Chetlur

ISBN: 978-3-031-01251-8   paperback
ISBN: 978-3-031-02379-8   ebook
ISBN: 978-3-031-00243-4   hardcover

DOI 10.1007/978-3-031-02379-8

A Publication in the Springer Nature series
SYNTHESIS LECTURES ON LEARNING, NETWORKS, AND ALGORITHMS

Lecture #24
Editor: Lei Ying, *University of Michigan, Ann Arbor*
Editor Emeritus: R. Srikant, *University of Illinois at Urbana–Champaign*
Founding Editor Emeritus: Jean Walrand, *University of California, Berkeley*
Series ISSN
Print 2690-4306   Electronic 2690-4314

Poisson Line Cox Process

Foundations and Applications to Vehicular Networks

Harpreet S. Dhillon and Vishnu Vardhan Chetlur
Virginia Tech

SYNTHESIS LECTURES ON LEARNING, NETWORKS, AND ALGORITHMS
#24

ABSTRACT

This book provides a comprehensive treatment of the Poisson line Cox process (PLCP) and its applications to vehicular networks. The PLCP is constructed by placing points on each line of a Poisson line process (PLP) as per an independent Poisson point process (PPP). For vehicular applications, one can imagine the layout of the road network as a PLP and the vehicles on the roads as the points of the PLCP. First, a brief historical account of the evolution of the theory of PLP is provided to familiarize readers with the seminal contributions in this area. In order to provide a self-contained treatment of this topic, the construction and key fundamental properties of both PLP and PLCP are discussed in detail. The rest of the book is devoted to the applications of these models to a variety of wireless networks, including vehicular communication networks and localization networks. Specifically, modeling the locations of vehicular nodes and roadside units (RSUs) using PLCP, the signal-to-interference-plus-noise ratio (SINR)-based coverage analysis is presented for both ad hoc and cellular network models. For a similar setting, the load on the cellular macro base stations (MBSs) and RSUs in a vehicular network is also characterized analytically. For the localization networks, PLP is used to model blockages, which is shown to facilitate the characterization of asymptotic blind spot probability in a localization application. Finally, the path distance characteristics for a special case of PLCP are analyzed, which can be leveraged to answer critical questions in the areas of transportation networks and urban planning. The book is concluded with concrete suggestions on future directions of research.

Based largely on the original research of the authors, this is the first book that specifically focuses on the self-contained mathematical treatment of the PLCP. The ideal audience of this book is graduate students as well as researchers in academia and industry who are familiar with probability theory, have some exposure to point processes, and are interested in the field of stochastic geometry and vehicular networks. Given the diverse backgrounds of the potential readers, the focus has been on providing an accessible and pedagogical treatment of this topic by consciously avoiding the measure theoretic details without compromising mathematical rigor.

KEYWORDS

stochastic geometry, Poisson line Cox process (PLCP), Poisson line process (PLP), coverage probability, vehicular networks, vehicular ad hoc network (VANET), cellular vehicle-to-everything (C-V2X)

To Harnaaz, Donia, and our beloved family.
-HSD

To my wonderful family.
-VVC

Contents

Preface

The last decade has witnessed a wide acceptance of stochastic geometry for the mathematical analysis of wireless networks. In parallel, there have been many technological advancements in the area of vehicular communications, which are pivotal for realizing the vision of a connected intelligent transportation system. This book is a product of the synergy of these two critical developments. In particular, this book presents a new stochastic geometry approach to the modeling and analysis of vehicular networks in which the road layout is modeled using the Poisson line process (PLP) and the vehicles on each road are modeled using an independent Poisson point process (PPP). The locations of the vehicles, therefore, form a doubly-stochastic point process, termed the Poisson line Cox process (PLCP). Based largely on our original research, this is the first self-contained book specifically focusing on the PLCP. By consciously avoiding the measure theoretic details, we offer a more accessible and pedagogical treatment of this topic with the only prerequisites being probability theory and some exposure to point processes. Consequently, this book is useful for graduate students and researchers from both academia and industry, who are interested in the fields of stochastic geometry and vehicular networks. It can also be used as a reference for a few lectures on line processes within an advanced graduate course on stochastic geometry and its applications.

As will be shown in Chapter 1, the contents of this book can be categorized into two parts. In the first part spanning Chapters 2–4, our aim is to present a self-contained and rigorous mathematical treatment of PLCP. In order to familiarize readers with the rich history of line processes, we first provide a brief account of the seminal works in this area in Chapter 1. In Chapter 2, we discuss the relevant properties of the PPP, which are useful throughout the book. Building on this, we discuss the construction and key properties of the PLP and the PLCP in Chapters 3 and 4, respectively. In the second part of the book, we first apply these tools to model and analyze various scenarios of vehicular networks in Chapters 5–8. Owing to their versatility and tractability, the applications of PLP and PLCP extend beyond vehicular networks. We explore the utility of PLP and PLCP to localization networks, transportation networks, and urban planning in Chapters 9 and 10. As the development of the theory of PLCP and its applications to vehicular networks are fairly recent, several exciting problems in this domain are yet to be explored. Some of the promising future directions of research are discussed in Chapter 11.

It is our sincere hope that the accessible style of this book will inspire and guide researchers to conceive, formulate, and solve useful problems in this exciting new area of research.

Harpreet S. Dhillon and Vishnu Vardhan Chetlur
June 2020

Acknowledgments

HSD is deeply indebted to Jeff Andrews, François Baccelli, Radha Krishna Ganti, and Martin Haenggi for many discussions and long-term collaborations, which have been instrumental in shaping his vision for research in the general area of stochastic geometry. They all have contributed to his progress in numerous ways for which he will always remain grateful. He would also like to thank Justin Coon and Carl Dettmann for recent collaborations and several fruitful discussions on the topic of line processes. Their gracious invitation to deliver a tutorial on this topic at Bristol made him and VVC think more generally about this topic, which ultimately led to this book.

The authors also express gratitude to Justin Coon, Carl Dettmann, and Martin Haenggi for their meticulous reading and constructive feedback, which significantly improved the quality of this work. In particular, their suggestion of adding a chapter on the Poisson point process to make this book self-contained improved the overall organization of the material. Chapters 9 and 10 of the book are based on the joint works with Sundar Aditya, Andy Molisch, and Carl Dettmann. The authors thank them for their feedback and inputs. HSD would also like to thank Mike Buehrer, Jeff Reed, and Walid Saad for many stimulating discussions during our collaborative projects that have collectively benefited this work.

The authors are thankful to R. Srikant and Lei Ying, Series Editors of the *Synthesis Lectures on Learning, Networks, and Algorithms*, for their encouragement to write this book. Thanks to Michael Morgan, President of Morgan & Claypool Publishers, for being so patient and supportive during the development of this book. Thanks are also due to Christine Kiilerich, Assistant Editor at M&C, for coordinating the publication process so efficiently, and to Brent Beckley, Direct Marketing Manager at M&C, and HSD's Ph.D. student Chiranjib Saha for designing the front cover.

Finally, the authors are grateful to the United States National Science Foundation for supporting their research through Grant IIS-1633363 that led to this book.

Harpreet S. Dhillon and Vishnu Vardhan Chetlur
June 2020

CHAPTER 1

Introduction

The design and performance analysis of large-scale wireless networks, such as cellular networks, have mostly relied on extensive system-level simulations. However, with growing complexity of wireless technologies and large number of network parameters, the simulation-based approaches to design are becoming more error-prone and time consuming. This necessitates the development of complementary analytical methods that could provide insights into the dependencies between the network performance and design parameters. One such method that has gained popularity in the last decade is the use of tools from stochastic geometry which has enabled the analytical characterization of several key performance metrics in wireless networks. Interested readers are advised to refer to textbooks and monographs [1–13] for a pedagogical treatment of this topic and to [14–18] for tutorials and surveys focusing on the applications of stochastic geometry to wireless networks.

The basic idea behind stochastic geometry based analysis of wireless networks is to model the locations of the nodes in a network as a point process and then characterize the performance metrics such as coverage and rate, which depend on the relative positions of the nodes, by leveraging the properties of these point processes, e.g., see [19–25] for a small sample of relevant papers. Due to its unparalleled tractability, the homogeneous 2D Poisson point process (PPP) is the natural first choice for modeling wireless networks. This is evident from its numerous applications to a variety of ad hoc and cellular network settings.[1] However, one of the defining properties of the PPP is its complete spatial randomness and hence it is not capable of capturing any spatial coupling between the location of nodes which is often the case in reality. Despite all its merits, the homogeneous PPP is still a single parameter model and is hence not suitable for capturing many spatial configurations of practical interest. The most relevant example in the context of this book is that of a vehicular communication network, which is discussed next.

A vehicular communication network is a heterogeneous network that consists of a variety of nodes distributed along the roadways and sidewalks, such as the vehicular nodes, roadside units (RSUs), pedestrians, and cyclists, as well as cellular macro base stations (MBSs) that are deployed across the region to provide ubiquitous coverage. As vehicular communication networks play a prominent role in several critical applications, such as autonomous driving and smart navigation, it is necessary to provide reliable and low latency wireless links between these nodes in a vehicular network. In order to gain reliable insights into the operation of a vehicu-

---

[1]Since this line of research is already mature, interested readers are advised to refer to books and tutorials, such as [6–18], for more details.

lar network, it is of paramount importance to consider spatial models that faithfully capture its distinct spatial geometry. Unlike most conventional wireless network settings, the locations of the nodes in a vehicular network are coupled with the underlying layout of the road system. This spatial coupling between the nodes cannot be captured by the homogeneous 2D PPP model. A few works in the literature have relied on simple linear stochastic models which represent multiple lanes of a road or an intersection of two roads. While this is a reasonable first-order model that is accurate in sparse scenarios, such as highways, it is too simplistic to capture the unique geometry of vehicular networks in denser environments, which are often more important from the design perspective. Therefore, we must consider spatial models that account for the randomness in the positions of the nodes on each road, as well as the irregularity in the spatial layout of roads.

Some of the candidate models that have been considered in the literature for modeling road networks include random planar graphs ranging from simple Erdős-Rényi (ER) graph [26, 27] to more sophisticated Watts-Strogatz and Barabási-Albert (BA) models [28–30]. While such models are often employed to study various topological properties of transportation networks, they are not well-suited for the analysis of communication networks. The set of spatial models that have been advocated for modeling road systems in the stochastic geometry literature are the edges of random tessellations such as Poisson Voronoi tessellation (PVT), Poisson Delaunay tessellation (PDT), and Poisson line tessellation (PLT) [31–34]. As it is useful to consider spatial models that are mathematically tractable, the PLT or Poisson line process (PLP) is often preferred over other models. Consequently, the Poisson line Cox process (PLCP) obtained by populating points on each line of the PLP as a 1D PPP has recently emerged as a popular choice for the modeling and analysis of vehicular networks [35–42].

1.1 MOTIVATION

While the PLCP model has been advocated for vehicular networks nearly two decades ago in [43], its mathematical treatment is fairly recent. This is partly because the primary interest of the community over the past two decades has been on developing the PPP-based models. Hence, unlike the case of PPPs, there is a scarcity of resources that provide a self-contained and rigorous mathematical treatment of the PLCP. While some of the basic properties of PLCP have been applied recently in the analyses presented in [35–42], it is never easy for the beginners in any field to get a formal understanding of the underlying theory from research papers. Therefore, this book is intended to serve as a reference for readers interested in learning the fundamentals of the PLCP and its applications to vehicular networks. The initial chapters of this book have been dedicated to an elaborate discussion on the theory of PPP, PLP, and PLCP for this purpose. The readers with background in measure theory are advised to refer to [2, 3], [4, Chapter 8], for a more detailed account on the theory of the PLPs. For completeness and to make these tools available to a wider audience (especially from the wireless networks area), we provide an accessible introduction to the key aspects of PLP that are necessary for the construction of the

PLCP and the discussion of its applications. A brief history of the theory of PLP and PLCP and its more recent applications in the literature are discussed next.

1.2 HISTORICAL PERSPECTIVE

In this section, we present a brief history of the theory of line processes and their relevance to communication networks. While the origin of problems pertaining to random lines in a plane can be traced back to the classical Buffon's needle problem (1777) [44], more concrete works that are pertinent to the modern theory of geometric probability began only in the mid-twentieth century. We will trace the evolution of the theory of line processes from S. Goudsmit's work in 1945 [45] and go through some of the famous works of R. E. Miles and M. S. Bartlett. We will then review some recent applications of line process to wireless communication systems.

In early 1940s, N. Bohr requested S. Goudsmit to investigate the probability of intersection of multiple independent tracks of subatomic particles at a single point for his cloud-chamber experiments. As an initial step toward the solution, Goudsmit considered a modified and simpler version of the problem in one of the first prominent papers on line processes published in 1945 [45]. Assuming a random set of lines that are parallel to the x and y axes, he characterized the area of an arbitrary rectangular fragment formed by the random lines. He further studied the general version of the problem by mapping the random lines to circles on a sphere in this paper.

Nearly twenty years after Goudsmit's paper, in 1964, R. E. Miles wrote two articles based on his doctoral thesis, where he presented several fundamental properties of line processes. He presented the distribution of the number of lines of a PLP hitting a convex region in the plane and also studied the key characteristics of the random polygons formed by the lines in the plane such as the number of sides (vertices), perimeter, area, and the diameter of the incircle of the convex polygons [46, 47]. In the same year, P. I. Richards wrote a paper with some additional results related to the averages of the random polygons by generalizing Goudsmit's approach [48]. Further, Miles wrote a series of papers where he studied various properties of line processes [49–52]. In 1973, he wrote a paper investigating the ergodic properties of the aggregates of the random polygons generated by the planar PLP [49]. A sequel to this paper presenting the results of a Monte Carlo study of random polygons was published in 1976 [50].

The theory of line processes appears to have gained more attention in the decade that followed the initial works of Miles. In 1963–1967, M. S. Bartlett focused on the empirical spectral analysis of point process and line process as a statistical method of studying the deviations from completely random processes [53–55]. Bartlett coined the term *line process* in his first paper on this topic. R. Davidson, along with D. Kendall, focused on setting up a theoretical framework for Bartlett's work as a part of his doctoral thesis [4]. In 1970, Davidson also published his work on the analysis of second-order properties of stationary line processes [56]. The development of the theory of line processes further extended into the study of stochastic processes of random objects in higher dimensions, such as plane processes and flat processes [4, 57]. L. A. Santaló and I. Yanez studied the properties of polygons formed by random lines in hyperbolic planes in [58].

The applications of line processes has spanned various fields, including material science, geology, image processing, transportation, localization, and wired and wireless communications [31–43, 59–65]. For instance, the line processes were used to model the positions of fibers in each layer of a fiber membrane in [59]. The strength of the fiber membrane is then analyzed by studying the pores of the fiber which are represented by the polygons formed by the lines of the line process. PLP has also been used in the modeling of fracture patterns on rocks and other surfaces [60]. In the area of image processing, PLT has been used to partition an image into multiple non-overlapping regions to analyze its statistical properties [61]. In 1997, F. Baccelli proposed to model road systems by a PLP to analyze the handover rate in cellular networks [43]. We will now mainly focus on some of the prominent works in the communications literature that have been motivated by this idea.

V. Schmidt and his coauthors wrote a series of papers exploring the idea of modeling road systems by PLT for the analysis of fixed access networks [31–34, 66]. In [31], the authors explored the idea of fitting various tessellation models (such as PVT, PDT, and PLT) to the infrastructure data of Paris. In [32], the authors modeled the cable trench system located along the roads by a PLP and the network components by a PPP, thereby forming a PLCP and analyzed the shortest path length along the roads between two network components located on the lines. In order to analyze the performance of wireless networks, the Euclidean distance between the points of a PLCP has been investigated in [33]. Further, in [34], the authors analyzed the geometric properties of the typical cell in a Voronoi tessellation generated by a PLCP.

While the layout of roads is modeled by the lines of a PLP, the polygons formed by the lines of the PLP can be visualized as blocks in urban or suburban areas. This idea has led to the use of PLPs to model the effect of blockages in wireless networks. The authors of [67] have analyzed the signal-to-interference-plus-noise ratio (SINR)-based coverage of the typical receiver in an urban cellular network where the shadowing effects are modeled using a simple variant of the PLP. The coverage of the typical receiver in an urban cellular network for millimeter-wave communications has been investigated in [65], where the dominant signal from the transmitter to the receiver is the one that travels along the road segments because of high penetration losses through buildings. The asymptotic blind spot probability in a localization network has been analyzed in [62] by modeling the obstacles as a PLP.

Due to the growing interest in the analysis of vehicular communication networks, the PLCP model has also been employed in several works to model the locations of vehicular nodes, RSUs, and other network infrastructure along the roads to characterize the performance of the network [35–41, 64, 68, 69]. Using this spatial model, F. Morlot derived the uplink coverage probability for a setup in which the locations of transmitter nodes are modeled by the PLCP [64]. The performance of a vehicular ad hoc network (VANET) has been studied using the PLCP in [38, 68]. The SINR-based coverage analysis of the typical receiver node in a vehicular network for the cellular network model has been presented in [35, 40]. The downlink rate coverage of the typical receiver has also been recently investigated in [37, 39]. The analytical results presented in some of these works will be discussed in detail in this book.

1.3 SCOPE AND ORGANIZATION

In this book, we provide a concise exposition of the theory of PLP and PLCP. We demonstrate the analytical characterization of several key performance metrics using PLCP for vehicular communication networks. In addition to that, we also show that the PLP and PLCP are quite useful in the modeling and analysis of localization and transportation networks. The scope and organization of this book are summarized as follows.

- In Chapter 2, we provide a brief introduction to the theory of PPPs and discuss some of the fundamental properties of the homogeneous PPP that will be useful in our subsequent discussion on PLCP.

- In Chapter 3, we explain the construction of planar line processes by establishing their relation to point processes. We specifically focus on PLP and some of its fundamental properties.

- In Chapter 4, we present the theory of PLCP which includes the construction of PLCP and some of its basic properties such as motion-invariance, void probability, nearest-neighbor distance distribution, and Palm distribution. We also present the asymptotic behavior of the PLCP, where we prove the convergence of a PLCP to homogeneous 1D and 2D PPPs under different conditions. These properties are quite useful in the performance analysis of wireless networks where the locations of the nodes are modeled by PLCP.

- We present an outline for the computation of SINR-based coverage probability in vehicular communication networks in Chapter 5. We discuss in detail the steps involved in the derivation of this result using the PLCP for the ad hoc and cellular network models in Chapters 6 and 7, respectively. We also present some numerical results and discuss key performance trends.

- In Chapter 8, we characterize the distribution of load on MBSs and RSUs in a vehicular network in which the locations of the vehicular users are modeled by PLCP.

- In Chapter 9, we present the application of PLP to localization networks. By modeling the obstacles in a network by the lines of PLP, we investigate the localizability of a desired target node.

- In Chapter 10, we consider a simple variant of the PLP and characterize the length of the shortest path between neighboring points in the sense of path distance. We then discuss the

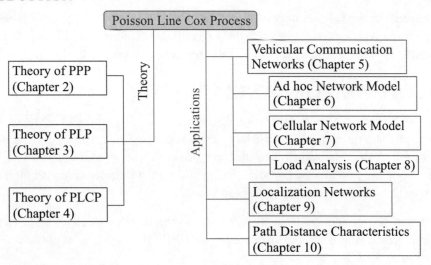

Figure 1.1: Organization of this book.

applications of such results in answering important questions pertaining to transportation networks and urban planning.

• In Chapter 11, we suggest some future directions of research from the perspective of spatial modeling, performance metrics, and mathematical analysis.

The organization of this book is also depicted in Fig. 1.1.

CHAPTER 2

The Poisson Point Process

In this chapter, we present a brief introduction to the theory of the PPP, which is one of the primary building blocks of the PLCP. We will also discuss the basic properties of the PVT generated by a homogeneous 2D PPP. As the main goal of this chapter is to facilitate the readers in understanding the theory of the PLCP in the subsequent chapters, we limit our focus to only those properties of PPPs that are relevant to this book. Interested readers are advised to refer to standard textbooks, such as [4, 6], for a more comprehensive discussion on this topic.

2.1   INTRODUCTION

The PPP is the most extensively studied point process which has found applications in various areas due to its unparalleled analytical tractability. The formal definition of the PPP is provided next.

Definition 2.1   (The Poisson point process.) A point process $\Phi \subset \mathbb{R}^d$ is a PPP if it exhibits the following two properties.

- *Poisson distributed point counts*: The number of points of Φ within a bounded Borel set $B \subset \mathbb{R}^d$ follows a Poisson distribution, i.e.,

$$\mathbb{P}(N(B) = n) = \frac{\exp(-\Lambda(B))(\Lambda(B))^n}{n!}, \tag{2.1}$$

  where Λ is the *intensity measure* which is defined as $\Lambda(B) = \mathbb{E}[N(B)]$.

- *Independent scattering*: If $B_1, B_2, \ldots$ are disjoint bounded Borel sets, then the number of points in those sets $N(B_1), N(B_2), \ldots$ are independent random variables.

While the above is a general definition of PPP, we are mainly interested in the homogeneous PPP in the context of this book.

Definition 2.2   (Homogeneous PPP.) A PPP Φ with a constant intensity λ is called the homogeneous PPP. Here, λ is the mean number of points per unit volume in $\mathbb{R}^d$.

A realization of the homogenous 2D PPP is depicted in Fig. 2.1.

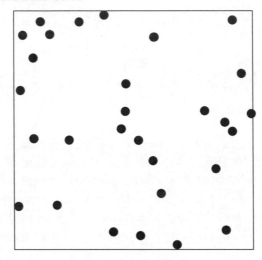

Figure 2.1: Illustration of the homogeneous PPP in $\mathbb{R}^2$.

2.2 PROPERTIES OF PPP

We will now study some basic properties of the PPP that will be relevant to our discussion on PLP and PLCP in the later chapters. As mentioned earlier, we will mostly focus on the homogeneous PPP. While we are specifically interested in a 2D homogeneous PPP, we will present most of the results in this section for a homogeneous PPP in $\mathbb{R}^d$ since the exposition is not too different for the two cases.

2.2.1 STATIONARITY AND MOTION-INVARIANCE

A point process is said to be *stationary* if its distribution is invariant to translation. As an example, let us consider a transformation $T_{(t,\gamma)}$ in the 2D case, which denotes the translation of the origin by a distance t in a direction that subtends an angle γ w.r.t. the positive x-axis in the anticlockwise direction. Upon applying this transformation on a 2D PPP $\Phi \equiv \{x_i\}$, the coordinates of these points change from (x, y) to $(x - t \cos \gamma, y - t \sin \gamma)$. Thus, the 2D PPP Φ is stationary iff the translated point process $T_{(t,\gamma)}\Phi$ has the same distribution as Φ.

A point process is said to be *isotropic* if the distribution of its points is invariant to the rotation of the axes. Considering again the 2D case, upon rotating the axes by an angle ω in counter clockwise direction, the coordinates of the points of a 2D PPP change from (x, y) to $(x \cos \omega + y \sin \omega, -x \sin \omega + y \cos \omega)$. Thus, a 2D PPP Φ is isotropic if the rotated point process $R_\omega \Phi$ has the same distribution as Φ. A point process is said to be *motion-invariant* if it is both stationary and isotropic.

For a homogeneous 2D PPP with intensity λ, the number of points lying in any bounded set B is Poisson distributed with mean $\lambda|B|$, where $|B|$ denotes the area of the region. Since the area $|B|$ is unaffected by the translation and rotation of the axes, the distribution of points remains unchanged and hence, the homogeneous PPP is motion-invariant.

2.2.2   SUPERPOSITION

We will now examine the point process resulting from the superposition of independent homogeneous PPPs in the following lemma.

Lemma 2.3   *The point process Φ obtained from the superposition of two independent homogeneous PPPs Φ_1 and Φ_2 with intensities λ_1 and λ_2, respectively, is also a homogeneous PPP with intensity $\lambda = \lambda_1 + \lambda_2$.*

Proof.   Let us examine the distribution of the number of points of Φ in a bounded Borel set B. The probability mass function (PMF) of the number of points is

$$
\mathbb{P}(N(\Phi \cap B) = n) = \mathbb{P}(N((\Phi_1 \cup \Phi_2) \cap B) = n)
$$

$$
= \sum_{k=0}^{n} \mathbb{P}(N(\Phi_1 \cap B) = k) \, \mathbb{P}(N((\Phi_1 \cup \Phi_2) \cap B) = n \mid N(\Phi_1 \cap B) = k)
$$

$$
= \sum_{k=0}^{n} \mathbb{P}(N(\Phi_1 \cap B) = k) \, \mathbb{P}(N(\Phi_2 \cap B) = n - k)
$$

$$
= \sum_{k=0}^{n} \frac{\exp(-\lambda_1 |B|)(\lambda_1 |B|)^k}{k!} \frac{\exp(-\lambda_2 |B|)(\lambda_2 |B|)^{n-k}}{(n-k)!}
$$

$$
= \frac{1}{n!} \exp(-\lambda_1 |B| - \lambda_2 |B|) \sum_{k=0}^{n} \frac{n!}{k!(n-k)!} (\lambda_1 |B|)^k (\lambda_2 |B|)^{n-k}
$$

$$
= \frac{\exp(-(\lambda_1 + \lambda_2)|B|) \, ((\lambda_1 + \lambda_2)|B|)^n}{n!}.
$$

This shows that Φ is also a homogeneous PPP with intensity $\lambda_1 + \lambda_2$. $\square$

This result can be directly extended to the superposition of any finite number of independent PPPs.

2.2.3   INDEPENDENT THINNING

The thinning of a point process is one of the basic transformations by which some of the points of the point process are removed based on certain rules, as illustrated in Fig. 2.2. If the removal of each point is independent of the other points, then this process is referred to as independent thinning. We will now discuss the application of this transformation to a homogeneous PPP.

Lemma 2.4   *For a homogeneous PPP Φ with intensity λ, if the probability that a point located at x is retained is given by $h : \mathbb{R}^d \to [0, 1]$, then the thinned point process is also a PPP with intensity $\lambda h(x)$.*

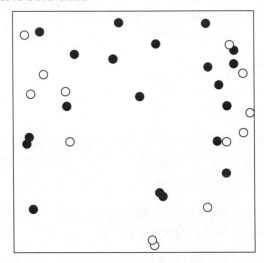

Figure 2.2: Illustration of a thinned PPP where the solid circles represent the points that were retained from the original PPP and the empty circles denote the points that were removed.

Proof. Let us denote the thinned point process by Φ'. The PMF of the number of points of Φ' lying inside a bounded Borel set B can be computed as

$$\mathbb{P}\left(N(\Phi' \cap B) = k\right) = \sum_{n=k}^{\infty} \mathbb{P}\left(N(\Phi \cap B) = n\right) \mathbb{P}\left(N(\Phi' \cap B) = k \mid N(\Phi \cap B) = n\right).$$

Before we proceed further, we will first determine the probability with which a point is retained after thinning. Since a point $x \in \Phi \cap B$ is uniformly distributed in B, the probability that this point is retained is computed as

$$\mathbb{P}(x \in (\Phi' \cap B) \mid x \in (\Phi \cap B)) = \frac{1}{|B|} \int_B h(x)dx, \tag{2.2}$$

where $|B|$ denotes the volume of B. Using this result, the PMF of the number of points of Φ' in B can be obtained as

$$\mathbb{P}\left(N(\Phi' \cap B) = k\right) = \sum_{n=k}^{\infty} \frac{e^{-\lambda|B|}(\lambda|B|)^n}{n!} \binom{n}{k} \left(\frac{1}{|B|}\int_B h(x)dx\right)^k \left(1 - \frac{1}{|B|}\int_B h(x)dx\right)^{n-k}$$

$$= \frac{e^{-\lambda|B|}(\lambda|B|)^k}{k!}\left(\frac{1}{|B|}\int_B h(x)dx\right)^k$$

$$\times \sum_{n=k}^{\infty} \frac{(\lambda|B|)^{n-k}}{(n-k)!}\left(1 - \frac{1}{|B|}\int_B h(x)dx\right)^{n-k}$$

$$= \frac{e^{-\lambda|B|}}{k!}\left(\lambda\int_B h(x)dx\right)^k \exp\left(\lambda|B| - \lambda\int_B h(x)dx\right)$$

$$= \frac{1}{k!}\exp\left(-\lambda\int_B h(x)dx\right)\left(\lambda\int_B h(x)dx\right)^k.$$

This proves that Φ' is also a PPP with intensity $\lambda h(x)$.   □

2.2.4   PALM DISTRIBUTION

It is sometimes necessary to characterize the probability of an event (equivalently, property of the point process) given that the point process contains a point at a specific location. This is termed as the Palm probability (or the Palm measure) of the point process. This concept is especially useful in wireless networks, where we are often interested in characterizing the performance of the network as observed from a specific node. Interested readers are advised to refer to [2, 4, 6, 7] for a rigorous introduction to this concept. The Palm probability of an event $\mathcal{A}$ given that the point process Φ has a point at $x \in \mathbb{R}^d$ is defined as:

$$\mathsf{P}_x(\mathcal{A}) \triangleq \mathbb{P}(\Phi \in \mathcal{A} \mid x \in \Phi). \tag{2.3}$$

In the above definition, the point $x \in \Phi$ is termed the typical point [4, 6]. For a stationary point process, we can consider the typical point as the origin without loss of generality. Therefore, for a homogeneous PPP Φ, the Palm probability can be written as

$$\mathsf{P}_o(\mathcal{A}) \triangleq \mathbb{P}(\Phi \in \mathcal{A} \mid o \in \Phi). \tag{2.4}$$

Recall that the point counts of the PPP in bounded disjoint sets are completely independent. As a result, conditioning on the event that a point is located at the origin does not affect the distribution of the rest of the points. Thus, we can write the Palm probability as

$$\mathsf{P}_o(\mathcal{A}) = \mathbb{P}(\Phi \cup \{o\} \in \mathcal{A}). \tag{2.5}$$

This means that conditioning on a point at the origin is equivalent to adding a point at the origin. This property of the homogeneous PPP is known as the *Slivnyak's theorem*. Throughout this chapter, whenever we consider the typical point to be located at the origin o, the underlying Palm distribution is implicitly understood.

Reduced Palm distribution. The reduced Palm distribution refers to the distribution of points excluding a point after conditioning on its location. For a homogeneous PPP Φ, the reduced Palm probability is

$$P_o^!(\mathcal{A}) \triangleq \mathbb{P}(\Phi \setminus \{o\} \in \mathcal{A} \mid o \in \Phi) = \mathbb{P}(\Phi \in \mathcal{A}). \tag{2.6}$$

Similar to the notation used for the Palm and reduced Palm probabilities above, we will denote the Palm expectation and the reduced Palm expectation by $\mathbb{E}_o$ and $\mathbb{E}_o^!$, respectively.

2.2.5   PROBABILITY GENERATING FUNCTIONAL

We will now present the probability generating functional (PGFL) of the PPP, which is an important result that will be used at several places in the subsequent chapters of the book. The formal definition of PGFL of a point process is provided next.

Definition 2.5   (PGFL.) Let $f : \mathbb{R}^d \to [0, 1]$ be a measurable function such that $1 - f(x)$ has a bounded support. The PGFL of the point process Φ is then defined as

$$G(f) \triangleq \mathbb{E}\left[\prod_{x \in \Phi} f(x)\right]. \tag{2.7}$$

The PGFL of the homogeneous PPP is derived in the following lemma.

Lemma 2.6   *The PGFL of a homogeneous PPP Φ with intensity λ is*

$$\mathbb{E}\left[\prod_{x \in \Phi} f(x)\right] = \exp\left(-\lambda \int_{\mathbb{R}^d} 1 - f(x)\mathrm{d}x\right). \tag{2.8}$$

Proof. The PGFL of Φ can be computed as

$$\mathbb{E}\left[\prod_{x \in \Phi} f(x)\right] = \lim_{r \to \infty} \mathbb{E}\left[\prod_{x \in \Phi \cap B(o,r)} f(x)\right],$$

where $B(o, r)$ denotes a d-dimensional ball of radius r centered at the origin. Upon conditioning on the number of points K inside this ball and taking the expectation w.r.t. K, the above equation can be written as

$$\mathbb{E}\left[\prod_{x \in \Phi} f(x)\right] = \lim_{r \to \infty} \mathbb{E}_K \mathbb{E}\left[\prod_{x \in \Phi \cap B(o,r)} f(x) \,\middle|\, N(\Phi \cap B(o,r)) = k\right].$$

Since we are conditioning on the number of points being k, each of them is uniformly and independently distributed in $B(o, r)$. Thus, we have

$$
\begin{aligned}
\mathbb{E}\left[\prod_{x \in \Phi} f(x)\right] &= \lim_{r \to \infty} \mathbb{E}_K\left[\left(\int_{B(o,r)} \frac{f(x)}{|B(o,r)|} dx\right)^k\right] \\
&= \lim_{r \to \infty} \sum_{k=0}^{\infty} \mathbb{P}\left(N(\Phi \cap B(o,r)) = k\right)\left(\int_{B(o,r)} \frac{f(x)}{|B(o,r)|} dx\right)^k \\
&= \lim_{r \to \infty} \sum_{k=0}^{\infty} \frac{\exp\left(-\lambda|B(o,r)|\right)\left(\lambda|B(o,r)|^k\right)}{k!}\left(\int_{B(o,r)} \frac{f(x)}{|B(o,r)|} dx\right)^k \\
&= \lim_{r \to \infty} \exp\left(-\lambda|B(o,r)|\right) \exp\left(\lambda \int_{B(o,r)} f(x) dx\right) \\
&= \lim_{r \to \infty} \exp\left(-\lambda \int_{B(o,r)} (1 - f(x)) dx\right) \\
&= \exp\left(-\lambda \int_{\mathbb{R}^d} (1 - f(x)) dx\right).
\end{aligned}
$$

This completes the proof. $\qquad\qquad\square$

2.3   THE POISSON VORONOI TESSELLATION

2.3.1   DEFINITIONS

We will first define the terms PVT, typical cell, and zero cell. While one can define these terms for a PPP in $\mathbb{R}^d$, we will confine this discussion to the 2D case, which is sufficient for the subsequent development of this book. Therefore, we will consider a homogeneous 2D PPP $\Phi \equiv \{x_i\}$ with intensity λ in the rest of this chapter.

Definition 2.7   (Poisson Voronoi tessellation.) A Poisson Voronoi cell with the nucleus at $x \in \Phi$ is defined as the locus of all the points in the space whose distance to the point x is smaller than or equal to the distance to any other point of the PPP Φ, which is mathematically expressed as

$$
V_x = \{y \in \mathbb{R}^2 : \|y - x\| \le \|y - x'\|, \forall x' \in \Phi\}, \tag{2.9}
$$

where $\|\cdot\|$ denotes the Euclidean norm. The aggregate of the Poisson Voronoi cells of all the points of the PPP Φ constitutes the PVT corresponding to Φ.

Definition 2.8   (Typical cell.) For the homogeneous PPP Φ, the Voronoi cell corresponding to the typical point $o \in \Phi$ is termed the typical cell and is defined as

$$
V_o = \{y \in \mathbb{R}^2 : \|y\| \le \|y - x'\|, \forall x' \in \Phi\}. \tag{2.10}
$$

Definition 2.9   (Zero cell.) The Voronoi cell that contains the origin o is called the zero cell. For the PPP Φ, the zero cell is

$$V(o) = \{y \in \mathbb{R}^2 : \|y - x\| \leq \|y - x'\|, \|x\| \leq \|x'\|, \forall x' \in \Phi\}. \tag{2.11}$$

2.3.2   MEAN AREAS

We will now present the mean areas of the typical cell and zero cell of a PVT in the following lemma.

Lemma 2.10   *The mean area of the typical cell and zero cell of the PVT generated by a homogeneous 2D PPP Φ with intensity λ are $\frac{1}{\lambda}$ and $\frac{1.280176}{\lambda}$, respectively.*

Proof. We will first present the Palm inversion formula which provides the relation between the distribution of a stationary point process and its Palm version [6, Theorem 8.3]. For a stationary point process Ψ with intensity λ and its typical cell V_o, we can write

$$\mathbb{E}[f(\Psi)] = \lambda \mathbb{E}_o \left[\int_{\mathbb{R}^2} f(\Psi_{-x}) \mathbf{1}(x \in V_o) dx \right], \tag{2.12}$$

where Ψ_{-x} is the point process Ψ translated by $-x$. Upon applying this result to the homogeneous 2D PPP Φ and taking $f(\Phi) = 1$, we have

$$1 = \lambda \mathbb{E}_o[|V_o|], \tag{2.13}$$

where $|V_o|$ denotes the area of the typical cell. Thus, we obtain the mean area of the typical cell as $\frac{1}{\lambda}$.

We will now compute the mean area of the zero cell. Using area-biased sampling, the probability density function (PDF) of the area of the zero cell can be obtained from the PDF of the area of the typical cell as [70]

$$f_{|V(o)|}(a) = \frac{a}{\int_0^\infty a f_{|V_o|}(a) da} f_{|V_o|}(a). \tag{2.14}$$

Thus, the mean area of the zero cell is related to mean area of the typical cell by

$$\mathbb{E}[|V(o)|] = \lambda \mathbb{E}_o \left[|V_o|^2 \right]. \tag{2.15}$$

In [71], the second moment of the area of the typical Poisson Voronoi cell is calculated as

$$\mathbb{E}_o \left[|V_o|^2 \right] \approx \frac{1.280176}{\lambda^2}. \tag{2.16}$$

Substituting (2.16) in (2.15), the mean area of the zero cell is obtained as $\frac{1.280176}{\lambda}$.   □

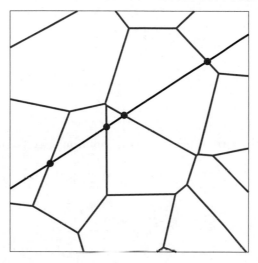

Figure 2.3: Illustration of the intersection points made by an arbitrary line crossing through the PVT.

2.3.3   LINE SECTION OF A PVT

In this subsection, we will characterize the intensity of the 1D point process formed by the intersection points of an arbitrary line with a PVT, as illustrated in Fig. 2.3. We will also present the PDF of the length of the chord segment in the typical Poisson Voronoi cell. In [71], Gilbert provided this result for a line crossing a PVT in a d-dimensional space. Since we do not need this result in the d-dimensional setting, we present a simpler proof for the more relevant 2D case provided in [72, 73] using the generalized version of the Buffon's needle argument. So, we will begin our discussion with the classical Buffon's needle problem in the following lemma.

Lemma 2.11   *The probability p_I with which a needle of length ℓ dropped on a planar array of parallel lines separated by a distance $d \geq \ell$ intersects at least one of the lines is $\frac{2\ell}{\pi d}$.*

Proof. First, let us consider an array of parallel lines as depicted in Fig. 2.4. We denote the angle subtended by the needle w.r.t. the direction of the lines by Θ which is uniformly distributed in the range $[0, \pi)$. For a given θ, the probability that the needle intersects a line is clearly the ratio of the length of the needle along the perpendicular direction to the lines to the separation between the lines, i.e.,

$$p_{I|\Theta} = \frac{\ell \sin \theta}{d}. \tag{2.17}$$

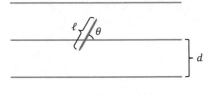

Figure 2.4: Illustration of the setting for the classical Buffon's needle problem.

Computing the expectation of the above expression w.r.t. Θ, we obtain

$$p_I = \frac{\ell}{\pi d} \int_0^\pi \sin\theta d\theta = \frac{2\ell}{\pi d}. \tag{2.18}$$

$\square$

We will now derive the mean edge length per unit area of a PVT. While this result has been presented for the 3D case in [72], we will derive this for the 2D case in the next lemma.

Lemma 2.12   *For a homogeneous 2D PPP Φ with density λ, the expected total length of edges per unit area of the corresponding PVT is $2\sqrt{\lambda}$.*

Proof. For the homogenous PPP Φ, consider the typical point $o \in \Phi$. We know that each edge of the PVT bounds two Poisson Voronoi cells and is located along the perpendicular bisector of the line connecting the corresponding nuclei. Therefore, the mean number of edges of the typical cell whose distances from the origin are between x and $x + dx$, denoted by $\mathbb{E}[N_e([x, x + dx])]$, is equal to the average number of points of the PPP that are located in the annular region centered at the origin with inner radius $2x$ and outer radius $2x + 2dx$. Thus, we have

$$\mathbb{E}[N_e([x, x + dx])] = \lambda \left(\pi(2x + 2dx)^2 - \pi(2x)^2 \right) = 8\pi\lambda x dx. \tag{2.19}$$

Let us consider an annular region centered at the origin with inner radius r and outer radius $r + dr$. As depicted in Fig. 2.5, for each edge E located at a distance $x < r$ from the origin; there are three possible events: (i) the edge E does not intersect the annular region; (ii) the edge E intersects the annular region once; and (iii) the edge E intersects the annular region at two points. We denote these events by $\mathcal{A}_0$, $\mathcal{A}_1$, and $\mathcal{A}_2$, respectively. The length of the edge inside the annular region in each of these cases is given by

$$\nu_1(E \cap \{B(o, r + dr) \setminus B(o, r)\}) = \begin{cases} 0, & \text{when } \mathcal{A}_0 \text{ occurs,} \\ \left(\sqrt{(r + dr)^2 - x^2} - \sqrt{r^2 - x^2} \right), & \text{when } \mathcal{A}_1 \text{ occurs,} \\ 2\left(\sqrt{(r + dr)^2 - x^2} - \sqrt{r^2 - x^2} \right), & \text{when } \mathcal{A}_2 \text{ occurs.} \end{cases} \tag{2.20}$$

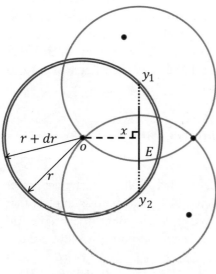

(a) $\mathcal{A}_0$: The edge does not intersect the annular region.

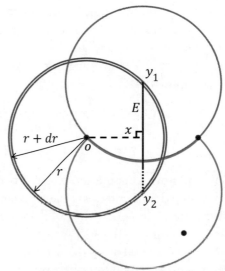

(b) $\mathcal{A}_1$: The edge intersects the annular region at one point.

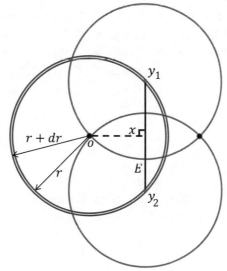

(c) $\mathcal{A}_2$: The edge intersects the annular region at two points.

Figure 2.5: Illustration for the proof of Lemma 2.12.

We will now compute the probability of occurrence of these events. We will first focus on the event $\mathcal{A}_2$. Let us denote the locations of the elements where the edge E intersects the annular region by y_1 and y_2, as shown in Fig. 2.5. If the edge E intersects the annular region at both these points, it implies that a circle of radius r centered at each of these locations passing through the origin and another point of the PPP does not contain any other point of the PPP inside it (refer to the orange circles in Fig. 2.5c). Thus, the probability of occurrence of $\mathcal{A}_2$ can be computed as

$$
\begin{aligned}
\mathbb{P}(\mathcal{A}_2) &= \mathbb{P}\left(N\left(\Phi \cap \{B(y_1, r) \cup B(y_2, r)\}\right) = 0\right) \\
&= \mathbb{P}\left(N\left(\Phi \cap B(y_1, r)\right) = 0\right) \mathbb{P}\left(N\left(\Phi \cap \{B(y_2, r) \setminus B(y_1, r)\}\right) = 0\right) \\
&= \exp\left(-\lambda \pi r^2 - \lambda |B(y_2, r) \setminus B(y_1, r)|\right).
\end{aligned}
\tag{2.21}
$$

We will now compute the probability of occurrence of $\mathcal{A}_1$. For this calculation, let us consider that the edge E intersects the annular region at y_1, as depicted in Fig. 2.5b. This implies that there are no points of the PPP inside the circle $B(y_1, r)$ that passes through the origin and another point of the PPP and simultaneously, there exists at least one point inside the circle $B(y_2, r)$ centered at y_2 (refer to the green circle in Fig. 2.5b). Note that we must also account for the other possibility where the edge E intersects the annular region at y_2 instead of y_1. Thus, the probability of occurrence of $\mathcal{A}_1$ can be computed as

$$
\begin{aligned}
\mathbb{P}(\mathcal{A}_1) &= \mathbb{P}\left(N\left(\Phi \cap B(y_1, r)\right) = 0, N\left(\Phi \cap B(y_2, r)\right) > 0\right) \\
&\quad + \mathbb{P}\left(N\left(\Phi \cap B(y_1, r)\right) > 0, N\left(\Phi \cap B(y_2, r)\right) = 0\right) \\
&= \mathbb{P}\left(N\left(\Phi \cap B(y_1, r)\right) = 0\right) \mathbb{P}\left(N\left(\Phi \cap \{B(y_2, r) \setminus B(y_1, r)\}\right) > 0\right) \\
&\quad + \mathbb{P}\left(N\left(\Phi \cap B(y_2, r)\right) = 0\right) \mathbb{P}\left(N\left(\Phi \cap \{B(y_1, r) \setminus B(y_2, r)\}\right) > 0\right) \\
&= 2\exp(-\lambda \pi r^2)\left(1 - \exp\left(-\lambda |B(y_2, r) \setminus B(y_1, r)|\right)\right).
\end{aligned}
\tag{2.22}
$$

Since the probability of events $\mathcal{A}_0$, $\mathcal{A}_1$, and $\mathcal{A}_2$ sum to 1, the probability of occurrence of $\mathcal{A}_0$ can be computed as

$$
\mathbb{P}(\mathcal{A}_0) = 1 - \mathbb{P}(\mathcal{A}_1) - \mathbb{P}(\mathcal{A}_2).
\tag{2.23}
$$

Using these results, the average length of the edge E located at a distance x from the origin inside the annular region $B(o, r + dr) \setminus B(o, r)$ is obtained as

$$
\begin{aligned}
\mathbb{E}\left[\nu_1(E \cap \{B(o, r + dr) \setminus B(o, r)\})\right] &= 0\mathbb{P}(\mathcal{A}_0) + \left(\sqrt{(r + dr)^2 - x^2} - \sqrt{r^2 - x^2}\right)\mathbb{P}(\mathcal{A}_1) \\
&\quad + 2\left(\sqrt{(r + dr)^2 - x^2} - \sqrt{r^2 - x^2}\right)\mathbb{P}(\mathcal{A}_2) \\
&= 2\left(\sqrt{(r + dr)^2 - x^2} - \sqrt{r^2 - x^2}\right)\exp(-\lambda \pi r^2).
\end{aligned}
\tag{2.24}
$$

We have now computed the expression for the average length of an edge in the annular region that is located at a distance x from the origin. From (2.19), we know the average number of edges that are located at distance x from the origin. Upon multiplying these expressions, we obtain the expression for the average of the total length of the edges in the annular region that are at a distance x from the origin. Further, upon integrating this expression from 0 to r, we obtain the average of the total length of the edges in the annular region as

$$\mathbb{E}[L_e(B(o, r + dr) \setminus B(o, r))] = \int_0^r 2\left(\sqrt{(r + dr)^2 - x^2} - \sqrt{r^2 - x^2}\right)\exp(-\lambda\pi r^2)8\pi\lambda x dx$$
$$= 16\lambda\pi r^2 \exp(-\lambda\pi r^2)dr. \tag{2.25}$$

Upon integrating the above expression, we obtain the mean perimeter of the typical cell as

$$\mathbb{E}[\nu_1(\partial V_0)] = \int_0^\infty 16\pi\lambda r^2 \exp(-\lambda\pi r^2)dr = \frac{4}{\sqrt{\lambda}}, \tag{2.26}$$

where ∂V_0 denotes the boundary of the typical cell.

As each edge is shared by a pair of points of the PPP, the mean edge length per unit area is obtained by multiplying the mean perimeter of the typical cell by $\frac{\lambda}{2}$. $\qquad\square$

Using this result, we now compute the mean number of intersections per unit length of a PVT with an arbitrary line. While this result can be derived using different methods proposed in the literature [71, 73, 74], we follow the technique presented in [73] in the next lemma.

Lemma 2.13 *The average number of intersection points per unit length on an arbitrary line drawn through a PVT corresponding to a homogeneous 2D PPP of intensity λ is $\frac{4\sqrt{\lambda}}{\pi}$.*

Proof. Let us consider a large square of side S in the 2D plane containing the PVT. We denote the average edge length per unit area of the PVT by τ. We now divide the square into very narrow thin strips of width δ and length S. Let us denote the average number of intersection points per unit length of an arbitrary line with the PVT by $\bar{N}$. Therefore, the average number of crossings in a strip is given by $\bar{N}S$ (since the length of the strip is S). We will now compute the average length of the edges of the PVT within a given strip at each crossing. Let us consider that an edge of the PVT intersects the strip at an angle Θ as depicted in Fig. 2.6. Now, the length of the edge inside the strip is given by

$$\ell = \frac{\delta}{\sin\Theta}, \tag{2.27}$$

where $\Theta \in [0, \pi)$. Note that Θ in the above expression is a random variable. In order to compute the average length of the edge in a strip at a crossing, we now need to determine the PDF of Θ conditioned on the event that the edge of the PVT intersects the strip. While this can be derived using integral geometry [75, 76], we use the setting of the classical Buffon's needle problem in Lemma 2.11 to obtain this result. Let us now consider a small linear element of the edge of

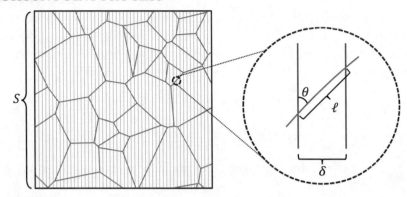

Figure 2.6: Illustration for the proof of Lemma 2.13.

length $d\ell$. We will now compute the PDF of the angle Θ conditioned on the event that this element intersects the edge of the strip. By Bayes' rule, the conditional PDF of Θ is given by

$$f_\Theta(\theta|I) = \frac{\mathbb{P}(I|\Theta)f_\Theta(\theta)}{\mathbb{P}(I)}, \tag{2.28}$$

where I denotes the event that the element of length $d\ell$ intersects the strip. From Lemma 2.11, we know that

$$\mathbb{P}(I|\theta) = \frac{d\ell \sin\theta}{\delta}, \tag{2.29}$$

$$\mathbb{P}(I) = \frac{2d\ell}{\pi\delta}, \tag{2.30}$$

$$f_\Theta(\theta) = \frac{1}{\pi}, \quad 0 \le \theta < \pi. \tag{2.31}$$

Substituting these expressions in (2.28), we obtain

$$f_\Theta(\theta|I) = \frac{1}{2}\sin\theta, \quad 0 \le \theta < \pi. \tag{2.32}$$

This result clearly makes sense because the probability of intersection of the element of the edge with the strip is directly proportional to the component normal to the strip. Using this result, the expected length of the edge within a strip per crossing can be computed as

$$\mathbb{E}[\ell] = \int_0^\pi \frac{\delta}{\sin\theta} f_\Theta(\theta|I)d\theta = \frac{\pi\delta}{2}. \tag{2.33}$$

As the total number of strips is $\frac{S}{\delta}$, the expected total length of the edges in the square can be computed as

$$\mathbb{E}[T] = \left(\frac{\pi\delta}{2}\right)(\bar{N}S)\left(\frac{S}{\delta}\right) = \frac{\pi\bar{N}S^2}{2}. \tag{2.34}$$

As the mean edge length per unit area is τ, we obtain

$$\bar{N} = \frac{2}{\pi}\tau.$$

From Lemma 2.12, we know that the average edge length per unit area of a PVT is $\tau = 2\sqrt{\lambda}$. Substituting this in the above equation, we obtain the desired result.   □

The inter-point distance of the point process formed by the intersection of the PVT with an arbitrary line is the same as the length of the chord segment of the line passing through the typical Poisson Voronoi cell. We provide the PDF of the length of this chord segment, denoted by C, in the following lemma.

Lemma 2.14   *Consider an arbitrary line passing through a PVT. The PDF of the length of the chord segment of this line in the typical cell, denoted by C above, is*

$$f_C(c) = \frac{\pi}{2}\lambda^{\frac{3}{2}} \int_0^\infty \int_0^\pi \tau \left[\lambda \left(\frac{\partial K_{\tau,\alpha}}{\partial c} \right)^2 - \frac{\partial^2 K_{\tau,\alpha}}{\partial c^2} \right] e^{-\lambda K_{\tau,\alpha}} \mathrm{d}\alpha \mathrm{d}\tau, \tag{2.35}$$

where

$$K_{\tau,\alpha} = 2\pi\tau^2 + \pi c^2 - 2\pi\tau c \cos\alpha - \tau^2 \left(\alpha - \frac{1}{2}\sin 2\alpha \right)$$
$$- (\tau^2 - 2\tau c \cos\alpha + c^2) \left(\phi - \frac{1}{2}\sin 2\phi \right), \tag{2.36}$$

and

$$\phi = \arccos \left[\frac{2c^2 - 2\tau c \cos\alpha}{2c\sqrt{\tau^2 - 2\tau c \cos\alpha + c^2}} \right]. \tag{2.37}$$

Proof. Let us denote the 1D point process formed by the intersection of an arbitrary directed line L with the PVT by Ξ_L. As mentioned earlier, the distribution of the chord length C in the typical cell is simply the inter-point distance distribution of the point process Ξ_L. From [4, 77], the relation between the cumulative distribution function (CDF) of the chord length of the line intersecting the typical cell and the linear contact distribution function $H_l(r)$ of Ξ_L is given by

$$H_l(r) = \frac{1}{m_C} \int_0^r (1 - F_C(c))\mathrm{d}c, \tag{2.38}$$

where m_C denotes the mean typical chord length. Upon differentiating the above equation w.r.t. r and applying Leibniz's rule, we obtain

$$F_C(r) = 1 - m_C \frac{\partial}{\partial r} H_l(r). \tag{2.39}$$

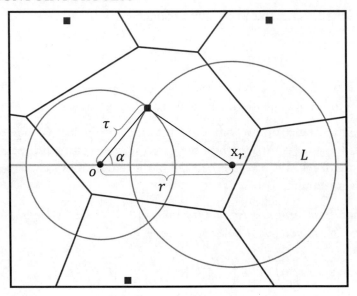

Figure 2.7: Illustration for the proof of Lemma 2.14.

From the intensity of the point process Ξ_L derived in Lemma 2.13, the mean typical chord length is given by

$$m_C = \frac{\pi}{4\sqrt{\lambda}}. \tag{2.40}$$

We will now focus on the computation of the linear contact distribution function of Ξ_L. For this calculation, we will align the line L along the x-axis without loss of generality. Now, the linear contact distribution can be defined as

$$H_l(r) = 1 - \mathbb{P}(N(\Xi_L \cap D_r) = 0), \tag{2.41}$$

where $D_r \equiv [o, x_r]$ denotes a line segment of length r with one endpoint at the origin o and the other endpoint at x_r in the direction of line L. We now need to compute the probability of the event $\mathcal{E}$ that there are no points of Ξ_L within the line segment $[o, r]$. We denote the polar coordinates of the nucleus of the cell containing the origin o by (τ, α), as shown in Figure 2.7. Since the point (τ, α) is the nearest point of the PPP Φ to the origin, we know that the PDF of τ is given by

$$f_\tau(\tau) = 2\pi\lambda\tau \exp(-\lambda\pi\tau^2), \tag{2.42}$$

and the angle α is uniformly distributed in the range $[0, 2\pi)$. Thus, conditioned on τ and α, the probability of occurrence of the event $\mathcal{E}$ is the probability that both o and x_r have the same nearest point of the PPP Φ. Mathematically, this means that there are no points of the PPP Φ inside the disc centered at x_r and radius $\sqrt{\tau^2 + r^2 - 2\tau r \cos\alpha}$, i.e.,

$$\mathbb{P}(\mathcal{E}|\tau,\alpha) = \mathbb{P}\left(N\left(\Phi \cap B(x_r, \sqrt{\tau^2 + r^2 - 2\tau r \cos\alpha})\right) = 0|\tau,\alpha\right)$$
$$\stackrel{(a)}{=} \exp\left(-\lambda \nu_2 \left(B(x_r, \sqrt{\tau^2 + r^2 - 2\tau r \cos\alpha}) \setminus B(o, \tau)\right)\right), \tag{2.43}$$

where (a) follows from the void probability of the 2D PPP and $\nu_2(\cdot)$ denotes the 2D Lebesgue measure (area). Upon computing the expectation of the above expression w.r.t. τ and α, we obtain the probability of occurrence of $\mathcal{E}$ as

$$\mathbb{P}(\mathcal{E}) = \int_0^{2\pi} \int_0^\infty \exp\left(-\lambda \nu_2 \left(B(x_r, \sqrt{\tau^2 + r^2 - 2\tau r \cos\alpha}) \setminus B(o, \tau)\right)\right) 2\pi \lambda \tau e^{-\lambda\pi\tau^2} d\tau \frac{d\alpha}{2\pi}$$
$$= \lambda \int_0^{2\pi} \int_0^\infty \exp\left(-\lambda \nu_2 \left(B(x_r, \sqrt{\tau^2 + r^2 - 2\tau r \cos\alpha}) \cup B(o, \tau)\right)\right) \tau d\tau d\alpha. \tag{2.44}$$

Upon substituting (2.44) and the expression for the area of the union of the two discs in (2.41), we obtain the linear contact distribution function as

$$H_l(r) = 1 - 2\lambda \int_0^\pi \int_0^\infty \exp\left(-\lambda K_{\tau,\alpha}\right) \tau d\tau d\alpha, \tag{2.45}$$

where

$$K_{\tau,\alpha} = 2\pi\tau^2 - 2\pi\tau r \cos\alpha - \tau^2\left(\alpha - \frac{1}{2}\sin 2\alpha\right)$$
$$+ \pi r^2 - (\tau^2 - 2\tau r \cos\alpha + r^2)\left(\phi - \frac{1}{2}\sin 2\phi\right), \tag{2.46}$$

and

$$\phi = \arccos\left[\frac{2r^2 - 2\tau r \cos\alpha}{2r\sqrt{\tau^2 - 2\tau r \cos\alpha + r^2}}\right]. \tag{2.47}$$

Substituting (2.40) and (2.45) in (2.39), we obtain the expression for the CDF of the typical chord length. We then obtain the PDF of the typical chord length by computing the derivative of $F_C(c)$ w.r.t. c. □

2.4   SUMMARY

In this chapter, we gave a brief introduction to the PPP with specific focus on some basic properties of homogeneous PPPs that will be useful in the later chapters of the book. We discussed the concepts of stationarity and motion-invariance and the operations on PPPs such as super-position and independent thinning. We also discussed the Palm distribution of PPPs, Slivnyak's theorem, and the PGFL for the homogeneous PPP. We then provided a brief account on some of the basic properties of the PVT. In particular, we presented the mean areas of the typical cell and the zero cell of the PVT. We then derived the intensity of the point process formed by the intersection of the PVT with an arbitrary line and presented the PDF of the chord length of the line passing through the typical Poisson Voronoi cell.

CHAPTER 3

The Poisson Line Process

In this chapter, we will first give a brief introduction to planar line processes. We will then specifically focus on PLP and some of its fundamental properties. Note that our intention here is to focus on the construction and specific aspects of PLP that would be useful in understanding some of the theoretical results pertaining to PLP-based spatial models and applications that will be discussed in the latter parts of this book.

3.1 PLANAR LINE PROCESSES

Planar line processes are a special case of flat processes, which are random systems of k-dimensional planes in $\mathbb{R}^d$. A planar line process is simply a random collection of lines in the 2D Euclidean plane. This prompts the question: *How does one endow a distribution on a set of lines?* The answer lies in the construction of the line process which is discussed next.

Any undirected line in $\mathbb{R}^2$ can be uniquely characterized by two parameters: (i) the perpendicular distance ρ of the line from the origin, and (ii) the angle θ subtended by the perpendicular dropped onto the line from the origin with respect to the positive x-axis in counterclockwise direction. The pair of parameters (ρ, θ) can be represented as the coordinates of a point on the cylindrical surface $\mathcal{C} \equiv \mathbb{R}^+ \times [0, 2\pi)$, which is termed the *representation space*. Therefore, each line in $\mathbb{R}^2$ can be uniquely mapped to a point in $\mathcal{C}$, as illustrated in Fig. 3.1. Thus, a random collection of lines can be constructed from a random set of points in $\mathcal{C}$. Such a collection of lines is referred to as a line process.

Note that the mapping between the lines in $\mathbb{R}^2$ and points in $\mathcal{C}$ provides us the flexibility to carry out the analysis in both the domains. As the theory of point processes is much more well-established, it is often preferable to study various aspects of line processes through the corresponding point process representation in $\mathcal{C}$. This technique will be employed at several places in this book. Also, this construction of line process is quite generic and some of the other methods of generation of line process can also be expressed using this construction. For instance, let us consider a line process in $\mathbb{R}^2$ constructed by drawing a line through each point of a 2D PPP in $\mathbb{R}^2$ such that the line is perpendicular to the line joining the point of the PPP to the origin. In this construction, each line is simply characterized by the corresponding point of the PPP. While this would be an infinitely dense line process, it is important to note that it can still be expressed as a point process in the representation space.

We will now discuss the concepts of stationarity, isotropy, and motion-invariance for planar line processes.

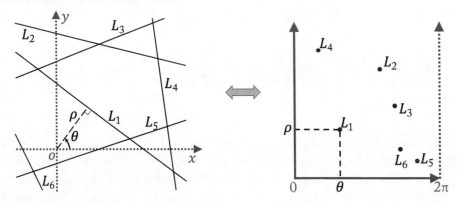

Figure 3.1: Illustration of the mapping between line process in $\mathbb{R}^2$ and point process in the representation space $\mathcal{C}$.

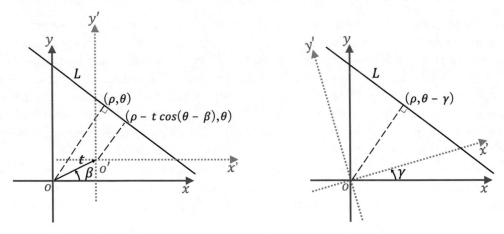

Figure 3.2: Illustration of the translation and rotation of the axes.

3.1.1   STATIONARITY

Analogous to point processes, a planar line process $\Psi_l = \{L_1, L_2, \dots\}$ is said to be stationary if its distribution is invariant to any translation in the plane. Let us consider a transformation $T_{(t,\beta)}$ corresponding to the translation of the origin by a distance t in a direction that subtends an angle β w.r.t. the positive x-axis in counter clockwise direction, as shown in Fig. 3.2. Upon applying this translation in $\mathbb{R}^2$, the coordinates of a point in $\mathcal{C}$ corresponding to a line L change from (ρ_L, θ_L) to $(\rho_L - t \cos(\theta_L - \beta), \theta_L)$. Thus, a line process Ψ_l is stationary iff the point process corresponding to the translated line process $T_{(t,\beta)}\Psi_l$ in the representation space $T_{(t,\beta)}\Psi_\mathcal{C} = \{(\rho_{L_1} - t \cos(\theta_{L_1} - \beta), \theta_{L_1}), (\rho_{L_2} - t \cos(\theta_{L_2} - \beta), \theta_{L_2}), \dots\}$ has the same distribution as the point process $\Psi_\mathcal{C} = \{(\rho_{L_1}, \theta_{L_1}), (\rho_{L_2}, \theta_{L_2}), \dots\}$.

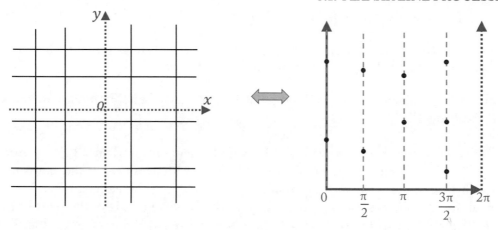

Figure 3.3: Illustration of the MPLP and the corresponding point process in the representation space.

3.1.2   ISOTROPY AND MOTION-INVARIANCE

A line process is said to be isotropic if the distribution of its lines is invariant under rotation of the axes. Upon rotating the axes about the origin by an angle γ w.r.t. the positive x-axis in anticlockwise direction, as shown in Fig. 3.2, the coordinates of the point corresponding to line L changes from (ρ_L, θ_L) to $(\rho_L, \theta_L - \gamma)$ in $\mathcal{C}$, where the operation $(\theta_L - \gamma)$ is modulo 2π. Thus, a line process is isotropic iff the point process corresponding to the rotated line process $R_\gamma \Psi_C = \{(\rho_{L_1}, \theta_{L_1} - \gamma), (\rho_{L_2}, \theta_{L_2} - \gamma), \dots\}$ has the same distribution as that of Ψ_C.

A line process is said to be motion-invariant if the distribution of lines is invariant to both translation and rotation of the axes. For example, let us consider a line process generated by a homogeneous 2D PPP Φ_C in $\mathcal{C}$. This line process is a PLP which will be defined shortly in the next section. Since Φ_C is stationary, the distribution of points in $\mathcal{C}$ remains unchanged under the translation and rotation of the axes in $\mathbb{R}^2$. Hence, the corresponding line process is motion-invariant. On the other hand, let us consider a line process generated by independent and homogeneous 1D PPPs along the lines $\theta = 0$, $\theta = \pi/2$, $\theta = \pi$, and $\theta = 3\pi/2$ in the representation space, as illustrated in Fig. 3.3. This would result in a set of horizontal and vertical lines in $\mathbb{R}^2$ called Manhattan Poisson line process (MPLP), which will be discussed in detail in Chapter 10. As the translation of the axes in $\mathbb{R}^2$ shifts the points in $\mathcal{C}$ only along the ρ-axis, the distribution of points in $\mathcal{C}$ remains unchanged due to the stationarity of the 1D PPPs. However, the rotation of the axes in $\mathbb{R}^2$ changes the distribution of the points in $\mathcal{C}$. Therefore, the MPLP is stationary but not isotropic.

3.2   PLP AND ITS PROPERTIES

Owing to its analytical tractability, the most popular line process model that has been studied in the literature is the PLP, which is formally defined as follows.

Definition 3.1   (Poisson line process.) The set of lines generated by a PPP in the representation space C is called a PLP.

Throughout this book, we will mostly consider motion-invariant PLPs. A motion-invariant PLP Φ_l is generated by a homogeneous PPP Φ_C with density λ_l in the representation space C. This can be easily verified by examining the conditions for stationarity and isotropy. First of all, we can observe that the determinant of the Jacobian matrix of the transformation from (ρ, θ) to $(\rho - t \cos(\theta - \beta), \theta)$ due to the translation $T_{(t,\beta)}$ is 1. Thus, this transformation does not alter the 2D Lebesgue measure (area) of any finite set in C. Also, note that this transformation shifts the points along ρ–axis and this shift is the same for all points with the same value of θ. As a result, the distribution of points in C is invariant to this transformation. Therefore, the distribution of points in $T_{(t,\beta)}\Phi_C$ is the same as that of Φ_C and hence, Φ_l is stationary. Also, it follows from the homogeneity of Φ_C that the orientation of the lines in Φ_l is uniformly distributed in the range $[0, 2\pi)$ and hence, Φ_l is isotropic. As Φ_l is both stationary and isotropic, it is motion-invariant. We will now discuss some of the fundamental properties of motion-invariant PLPs.

3.2.1   LINE DENSITY

The line density of a line process is defined as the mean line length per unit area. The relation between the line density of a motion-invariant PLP and the density of the corresponding point process in C is given in the following lemma.

Lemma 3.2   *For a PLP Φ_l generated by a homogeneous 2D PPP Φ_C with density λ_l in C, the line density μ_l is given by $\mu_l = \pi\lambda_l$.*

Proof. This result can be derived by computing the mean length of line segments of the PLP Φ_l inside any planar Borel set. Without loss of generality, let us consider a disc of unit radius centered at the origin $B(o, 1)$. By the definition of line density, we have

$$\mu_l = \frac{1}{\pi}\mathbb{E}\left[\sum_{L \in \Phi_l} \nu_1(L \cap B(o, 1))\right], \tag{3.1}$$

where $\nu_1(\cdot)$ denotes the 1D Lebesgue measure (length). For a line L at a distance ρ from the center of the disc, the length of the line segment inside the disc is $2\sqrt{1 - \rho^2}$ if $\rho \leq 1$. Further,

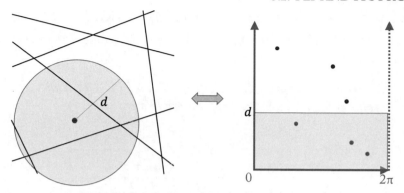

Figure 3.4: Illustration of the intersection of the lines of the PLP with a disc of radius d and its corresponding mapping in the representation space C.

rewriting the above equation in terms of the point process in C, we obtain

$$\mu_l = \frac{1}{\pi}\mathbb{E}\left[\sum_{(\rho,\theta)\in\Phi_C:\rho\leq 1} 2\sqrt{1-\rho^2}\right]. \tag{3.2}$$

By applying Campbell's theorem for sums over stationary processes [6], the above expression can be simplified as

$$\mu_l = \frac{1}{\pi}\lambda_l\int_0^{2\pi}\int_0^1 2\sqrt{1-\rho^2}\mathrm{d}\rho\mathrm{d}\theta = \lambda_l\pi. \tag{3.3}$$

This completes the proof. $\qquad\square$

3.2.2 LINES INTERSECTING A CONVEX REGION

The number of lines N_l of a motion-invariant PLP Φ_l with line density μ_l that intersect a convex region $K \subset \mathbb{R}^2$ follows a Poisson distribution with mean $\frac{\mu_l}{\pi}\nu_1(\partial K)$, where $\nu_1(\partial K)$ is the perimeter of the region K. This result follows from the distribution of points formed by the intersection of lines of a PLP with a curve of finite length, which was derived in [75] using integral geometry. For simplicity, we will show this result for a disc $B(o, d)$ of radius d centered at the origin. The number of lines of Φ_l that intersect the disc $B(o, d)$ is the same as the number of points of Φ_C that are inside the corresponding region of the disc in C. The disc $B(o, d)$ is represented in C by the rectangular region $B_C = \{\rho \leq d, 0 \leq \theta < 2\pi\}$, as shown in Fig. 3.4. As Φ_C is a homogeneous PPP with density λ_l, the number of points inside the rectangular region B_C follows a Poisson distribution with mean $\lambda_l(2\pi d) = \frac{\mu_l}{\pi}(2\pi d)$, where $2\pi d$ is the perimeter of the disc $B(o, d)$.

3.2.3 DISTANCE DISTRIBUTION OF THE nth CLOSEST LINE

We will now present the CDF and the PDF of the distance to the nth closest line from the origin, denoted by Y_n.

Lemma 3.3 *For a motion-invariant PLP Φ_l generated by a PPP with density λ_l in $\mathcal{C}$, the CDF and PDF of the distance of the nth closest line from the origin are*

$$CDF: \quad F_{Y_n}(y_n) = 1 - e^{-2\pi\lambda_l y_n} \sum_{k=0}^{n-1} \frac{(2\pi\lambda_l y_n)^k}{k!}, \tag{3.4}$$

$$PDF: \quad f_{Y_n}(y_n) = \frac{e^{-2\pi\lambda_l y_n}(2\pi\lambda_l y_n)^n}{y_n(n-1)!}. \tag{3.5}$$

Proof. By definition, the CDF of Y_n is

$$F_{Y_n}(y_n) = \mathbb{P}(Y_n \leq y_n) = 1 - \mathbb{P}(Y_n > y_n).$$

The event that the distance of the nth closest line Y_n is greater than y_n means that the number of lines that intersect the disc $B(o, y_n)$ centered at the origin with radius y_n would be at most $n-1$. Thus, we have

$$F_{Y_n}(y_n) = 1 - \mathbb{P}\left(N_l\big(B(o, y_n)\big) \leq n-1\right)$$

$$= 1 - \sum_{k=0}^{n-1} \mathbb{P}\left(N_l\big(B(o, y_n)\big) = k\right).$$

We know that the number of lines of Φ_l intersecting a disc of radius y_n is Poisson distributed with mean $2\pi\lambda_l y_n$. Thus, the CDF of Y_n is given by

$$F_{Y_n}(y_n) = 1 - \sum_{k=0}^{n-1} \frac{\exp(-2\pi\lambda_l y_n)(2\pi\lambda_l y_n)^k}{k!}.$$

The PDF $f_{Y_n}(y_n)$ can be obtained by computing the derivative of $F_{Y_n}(y_n)$ w.r.t. y_n. This completes the proof. $\square$

3.3 SUMMARY

In this chapter, we explained the construction of planar line processes in $\mathbb{R}^2$. We discussed the mapping between a line process in $\mathbb{R}^2$ and the corresponding point process in the representation space $\mathcal{C}$. We also discussed the criteria for stationarity, isotropy, and motion-invariance for line processes. We then particularly focused on the construction of a motion-invariant PLP and some of its fundamental properties, such as the line density, the distribution of number of lines intersecting a convex region, and the distance distribution of the nth closest line from the origin.

CHAPTER 4

The Poisson Line Cox Process

In this chapter, we will first present the construction of the PLCP. We will then study some of the fundamental properties of PLCP that will be useful in the later chapters for the analysis of vehicular networks modeled using this point process.

4.1 CONSTRUCTION OF PLCP

A doubly stochastic Poisson process or a Cox process is a Poisson process with a random intensity measure. In a PLCP, the source of randomness in the intensity measure of the point process is the underlying PLP. In order to understand this more clearly, we will first discuss the construction of a PLCP.

A PLCP, as illustrated in Fig. 4.1, is constructed by populating points on the lines of PLP such that the locations of points on each line form an independent 1D PPP. In this book, we limit our discussion to a PLCP $\Phi_p \equiv \cup_{L \in \Phi_l} \Psi_L$ in which the underlying PLP Φ_l is motion-invariant with line density μ_l and the locations of points on each line L form a homogeneous 1D PPP Ψ_L with density λ_p. Conditioned on the PLP Φ_l, the intensity measure of Φ_p is given by

$$\Lambda(A) = \mathbb{E}\left[N_p(A) \mid \Phi_l\right] = \lambda_p \sum_{L \in \Phi_l} \nu_1(L \cap A), \tag{4.1}$$

where $N_p(A)$ denotes the number of points of Φ_p in set $A \subset \mathbb{R}^2$. Let $\ell_A = \sum_{L \in \Phi_l} \nu_1(L \cap A)$. Note that ℓ_A, which represents the total length of line segments inside A, is random due to the underlying PLP.

4.2 PROPERTIES OF PLCP

In this section, we will study some basic properties of the PLCP such as stationarity, isotropy, void probability, Palm distribution, contact distance, and nearest-neighbor distance distributions. These properties will facilitate the analysis of various performance metrics in the following chapters of the book.

4.2.1 STATIONARITY AND MOTION-INVARIANCE

We know that a point process is stationary if the distribution of the points is invariant to translation. From the construction of PLCP, it follows that the number of points of the PLCP that

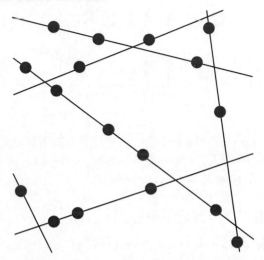

Figure 4.1: Illustration of the PLCP.

lie inside a set A follows a Poisson distribution with mean $\lambda_p \ell_A$. Now, let us consider that the origin is translated by a distance t in some random direction. Owing to the stationarity of the PLP Φ_l, the distribution of ℓ_A remains unchanged under this translation. Therefore, the distribution of $N_p(A)$ is also invariant to translation and, hence, Φ_p is stationary. Similarly, we can argue that Φ_p is isotropic due to the isotropy of the underlying PLP Φ_l. Therefore, the PLCP Φ_p is motion-invariant.

4.2.2   VOID PROBABILITY

As the distribution of points in a point process is completely specified by its void probability, we compute this for the PLCP Φ_p in the following lemma.

Lemma 4.1   *The void probability of the PLCP is*

$$\mathbb{P}\left(N_p(A) = 0\right) = \exp\left[-\lambda_l \int_0^{2\pi} \int_{\mathbb{R}^+} \left[1 - \exp\left(-\lambda_p \nu_1\left(L_{(\rho,\theta)} \cap A\right)\right)\right]\mathrm{d}\rho\mathrm{d}\theta\right], \quad (4.2)$$

where $A \subset \mathbb{R}^2$ is a Borel set.

Proof. As the locations of the points of the PLCP are restricted to the lines of the underlying PLP, the void probability can be computed as

$$\mathbb{P}\left(N_p(A) = 0\right) = \mathbb{P}\left(N_p\left(\cup_{L \in \Phi_l}\{L \cap A\}\right) = 0\right).$$

We now need to compute the probability that there are no points on any of the line segments inside the set A. We know that the number of points on any line is independent of the number of points on the other lines. Therefore, by conditioning on the underlying line process Φ_l, the void probability can be computed as

$$\mathbb{P}\left(N_p(A) = 0\right) = \mathbb{E}_{\Phi_l}\left[\prod_{L \in \Phi_l} \mathbb{P}\left(N_p(L \cap A) = 0 \mid \Phi_l\right)\right].$$

Using the void probability of 1D PPP on each line, we have

$$\mathbb{P}\left(N_p(A) = 0\right) = \mathbb{E}_{\Phi_l}\left[\prod_{L \in \Phi_l} \exp\left(-\lambda_p \nu_1\left(L \cap A\right)\right)\right].$$

By rewriting the above expression in terms of the point process Φ_c in the representation space, we obtain the void probability as

$$\mathbb{P}\left(N_p(A) = 0\right) = \mathbb{E}_{\Phi_c}\left[\prod_{(\rho,\theta) \in \Phi_c} \exp\left(-\lambda_p \nu_1\left(L_{(\rho,\theta)} \cap A\right)\right)\right],$$

where $L_{(\rho,\theta)}$ denotes a line with parameters (ρ, θ). Using the PGFL of the 2D PPP Φ_c, we obtain the final expression for the void probability as

$$\mathbb{P}\left(N_p(A) = 0\right) = \exp\left[-\lambda_l \int_0^{2\pi} \int_{\mathbb{R}^+}\left[1 - \exp\left(-\lambda_p \nu_1\left(L_{(\rho,\theta)} \cap A\right)\right)\right]d\rho d\theta\right].$$

$\square$

4.2.3 PALM DISTRIBUTION

As discussed already in Section 2.2.4, it is often necessary to consider the Palm distribution of the point process in many applications. Recall that the Palm probability of an event $\mathcal{A}$ given that the point process Φ_p has a point at x is defined as:

$$\mathsf{P}_x(\mathcal{A}) \triangleq \mathbb{P}(\Phi_p \in \mathcal{A} \mid x \in \Phi_p). \tag{4.3}$$

Since Φ_p is stationary, its conditional distribution $\mathsf{P}_x(\cdot)$ as seen from x given $x \in \Phi_p$ is the same as its conditional distribution $\mathsf{P}_o(\cdot)$ as seen from the origin o given $o \in \Phi_p$. In this context, $\mathsf{P}_o(\cdot)$ is also interpreted as the distribution of Φ_p as seen from its typical point at o. As the points of the PLCP lie on a line, there must exist a line L_0 of the underlying line process that passes through the origin. This means that there exists a point in Φ_c, corresponding to the line L_0, on the horizontal axis ($\rho = 0$) of the representation space $\mathcal{C}$. From Slivnyak's theorem [6], we know that the conditioning on the location of a point in a PPP is the same as adding that point to the PPP. Thus, by applying Slivnyak's theorem in the representation space, the resulting line process under the Palm probability of the PLCP is $\Phi_{l_0} \equiv \Phi_l \cup \{L_0\}$. Similarly, the point process on the line L_0 is the superposition of the 1D PPP Ψ_{L_0} with density λ_p and an atom at the origin, as a result of Slivnyak's theorem. Therefore, the Palm probability of the PLCP Φ_p is

$$P_o(\mathcal{A}) = \mathbb{P}(\Phi_p \in \mathcal{A} \mid o \in \Phi_p) = \mathbb{P}(\{\Phi_p \cup \Psi_{L_0} \cup \{o\}\} \in \mathcal{A}). \tag{4.4}$$

Thus, under the Palm distribution of the PLCP, the resulting point process Φ_{po} is the superposition of Φ_p, an independent 1D PPP with density λ_p on the line L_0, and an atom at the origin, i.e., $\Phi_{po} = \Phi_p \cup \Psi_{L_0} \cup \{o\}$.

Therefore, under the reduced Palm distribution of the PLCP, the resulting point process is given by $\Phi_{po}^! = \Phi_p \cup \Psi_{L_0}$.

Corollary 4.2   Under the reduced Palm distribution, the void probability of the PLCP is

$$P_o^! \left(N_p(A) = 0\right) = \exp\left[-\lambda_p \nu_1(L_0 \cap A)\right.$$
$$\left. -\lambda_l \int_0^{2\pi} \int_{\mathbb{R}^+} \left[1 - \exp\left(-\lambda_p \nu_1\left(L_{(\rho,\theta)} \cap A\right)\right)\right] \mathrm{d}\rho \mathrm{d}\theta\right]. \tag{4.5}$$

Proof. The proof follows along the same lines as that of Lemma 4.1 except that we must now include the line L_0 passing through the origin.   □

4.2.4   DISTRIBUTION OF POINTS IN A DISC

In this section, we will derive the PMF of the number of points of Φ_p in a disc of radius d. As Φ_p is motion-invariant, without loss of generality, we consider the disc $B(o, d)$ centered at the origin. Since the locations of points are restricted to the lines of the PLP Φ_l, in order to derive the PMF of the number of points, we need to characterize the total length of line segments inside $B(o, d)$, which will henceforth be referred to as the *total chord length*. So, we first derive the Laplace transform of the distribution of the total chord length T in the following lemma.

Lemma 4.3   *The Laplace transform of the total chord length distribution in $B(o, d)$ is*

$$\mathcal{L}_T(s) = \exp\left[-2\pi\lambda_l \int_0^d 1 - \exp(-2s\sqrt{d^2 - \rho^2})\mathrm{d}\rho\right]. \tag{4.6}$$

Proof. The Laplace transform of the distribution of the total chord length T in $B(o, d)$ can be computed as

$$\mathcal{L}_T(s) = \mathbb{E}\left[\exp\left(-s \sum_{L \in \Phi_l} \nu_1(B(o, d) \cap L)\right)\right].$$

For a line L at a distance $\rho < d$ from the origin, the length of the chord in $B(o, d)$ is $2\sqrt{d^2 - \rho^2}$. Thus, the above equation can be written as

$$\mathcal{L}_T(s) = \mathbb{E}\left[\prod_{\substack{(\rho,\theta)\in\Phi_c \\ \rho<d}} \exp\left(-2s\sqrt{d^2 - \rho^2}\right)\right].$$

Upon applying the PGFL for the 2D PPP Φ_c, we obtain the final expression for the Laplace transform of the total chord length distribution as

$$\mathcal{L}_T(s) = \exp\left[-\lambda_l \int_0^{2\pi}\int_0^d 1 - \exp(-2s\sqrt{d^2 - \rho^2})\mathrm{d}\rho\mathrm{d}\theta\right]$$

$$= \exp\left[-2\pi\lambda_l\int_0^d 1 - \exp(-2s\sqrt{d^2 - \rho^2})\mathrm{d}\rho\right].$$

$\square$

Using this result, we will now derive the PMF of the number of points of Φ_p in $B(o, d)$ in the following theorem.

Lemma 4.4   *The PMF of the number of points of the PLCP Φ_p in $B(o, d)$ is*

$$\mathbb{P}(N_p(B(o, d)) = k) = \frac{(-\lambda_p)^k}{k!}\left[\frac{\partial^k}{\partial s^k}\mathcal{L}_T(s)\right]_{s=\lambda_p}, \tag{4.7}$$

where $\mathcal{L}_T(s)$ is given in Lemma 4.3.

Proof. We know that the number of points of Φ_p on a line L that intersects $B(o, d)$ follows a Poisson distribution with mean $\lambda_p \nu_1(L \cap B(o, d))$. Since the number of points on each line is independent of the other lines, the total number of points of Φ_p in $B(o, d)$ follows a Poisson distribution with mean $\lambda_p T$, where $T = \sum_{L\in\Phi_l} \nu_1(B(o, d) \cap L)$ is the total chord length in $B(o, d)$. Thus, upon conditioning on T, the PMF of the number of points is

$$\mathbb{P}(N_p(B(o, d)) = k \mid T) = \frac{\exp(-\lambda_p t)(\lambda_p t)^k}{k!}.$$

Taking expectation of the above expression w.r.t. T, we obtain the overall PMF of the number of points of Φ_p in $B(o, d)$ as

$$\mathbb{P}(N_p(B(o, d)) = k) = \int_0^\infty \frac{\exp(-\lambda_p t)(\lambda_p t)^k}{k!} f_T(t)\mathrm{d}t.$$

Using the derivatives of the exponential function, the above expression can be rewritten as

$$\mathbb{P}(N_p(B(o,d)) = k) = \int_0^\infty \frac{(-\lambda_p)^k}{k!} \left[\frac{\partial^k e^{-st}}{\partial s^k} \right]_{s=\lambda_p} f_T(t) \mathrm{d}t.$$

Using the Leibniz's rule for the derivative of an integral in the above equation, we obtain

$$\mathbb{P}(N_p(B(o,d)) = k) = \frac{(-\lambda_p)^k}{k!} \left[\frac{\partial^k}{\partial s^k} \int_0^\infty e^{-st} f_T(t) \mathrm{d}t \right]_{s=\lambda_p}.$$

Note that the integral in the above equation is nothing but the Laplace transform of the total chord length distribution. Thus, the PMF of the number of points of Φ_p in $B(o,d)$ can be expressed in terms of $\mathcal{L}_T(s)$ as

$$\mathbb{P}(N_p(B(o,d)) = k) = \frac{(-\lambda_p)^k}{k!} \left[\frac{\partial^k}{\partial s^k} \mathcal{L}_T(s) \right]_{s=\lambda_p}.$$

$\square$

4.2.5   DISTRIBUTION OF POINTS IN A POISSON VORONOI CELL

While the PLCP is a suitable model for the locations of vehicular nodes, it is also necessary to understand the interaction between PLCP and other models such as a PVT which are often used to represent the coverage regions of MBSs in the network. This will be particularly useful in the load analysis of vehicular users which will be discussed in detail in Chapter 8. Therefore, in this subsection, we study the distribution of points of the PLCP in the typical cell of a PVT generated by a homogeneous 2D PPP Ω with density η.

As shown in the previous subsection, the first step is to compute the Laplace transform of the total chord length distribution in the typical cell V_o, which is given in the following lemma.

Lemma 4.5   *The Laplace transform of the total chord length distribution in V_o is*

$$\mathcal{L}_T(s) = 0.00086\eta^{3.7903} \int_0^\infty u^{6.5806} \exp\left[-\lambda_l u \int_0^\infty (1 - e^{-sc}) f_C(c) \mathrm{d}c - 0.1165(\sqrt{\eta}u)^{2.3361} \right] \mathrm{d}u, \tag{4.8}$$

where $f_C(c)$ is given in Lemma 2.14.

Proof.   Let us denote the chord length of the line L_i that intersects V_o by C_i. We also know that the number of lines N_l that intersect the typical cell V_o is random. Thus, upon conditioning on N_l, the Laplace transform of the chord length distribution can be computed as

$$\mathcal{L}_T(s \mid N_l) = \mathbb{E}\left[\exp\left(-s \sum_{i=1}^n C_i \right) \,\middle|\, N_l = n \right].$$

As the chord lengths corresponding to different lines that intersect V_o are independently and identically distributed, the above equation can be simplified as

$$\mathcal{L}_T(s \mid N_l) = \left[\int_0^\infty \exp(-sc) f_C(c) dc \right]^n,$$

where $f_C(c)$ is given by Lemma 2.14. From Section 3.2.2, recall that the number of lines of Φ_l that intersect a convex region $A \subset \mathbb{R}^2$ follows a Poisson distribution with mean $\lambda_l \nu_1(\partial A)$, where $\nu_1(\partial A)$ denotes the perimeter of A. Thus, upon conditioning on the perimeter U of the typical cell and taking the expectation w.r.t. N_l, we obtain

$$\mathcal{L}_T(s \mid U) = \sum_{n=0}^\infty \mathbb{P}(N_l = n | U) \left(\int_0^\infty \exp(-sc) f_C(c) dc \right)^n$$

$$= \sum_{n=0}^\infty \frac{\exp(-\lambda_l u)(\lambda_l u)^n}{n!} \left(\int_0^\infty \exp(-sc) f_C(c) dc \right)^n.$$

Using the Taylor series expansion for the exponential function, the above expression can be simplified as

$$\mathcal{L}_T(s \mid U) = \exp \left[-\lambda_l u + \lambda_l u \int_0^\infty \exp(-sc) f_C(c) dc \right], \qquad (4.9)$$

From [78], the empirical PDF of the perimeter of the typical cell is given by

$$f_U(u) = \frac{a b^{\frac{c}{a}}}{\Gamma\left(\frac{c}{a}\right)} \left(\frac{\sqrt{\eta}}{4} \right)^c u^{c-1} \exp \left(-b \left(\frac{\sqrt{\eta} u}{4} \right)^a \right), \qquad (4.10)$$

where $a = 2.33609$, $b = 2.97006$, and $c = 7.58060$. Upon computing the expectation of the expression in (4.9) w.r.t. U, we obtain the final expression. □

Lemma 4.6   *The PMF of the number of points of the PLCP Φ_p in the typical cell V_o is*

$$\mathbb{P}(N_p(V_o) = k) = \frac{(-\lambda_p)^k}{k!} \left[\frac{\partial^k}{\partial s^k} \mathcal{L}_T(s) \right]_{s=\lambda_p}, \qquad (4.11)$$

where $\mathcal{L}_T(s)$ is given in Lemma 4.5.

Proof. The proof follows along the same lines as that of Lemma 4.4. □

4.2.6   SPHERICAL CONTACT DISTRIBUTION FUNCTION

We will now determine the spherical contact distribution function of Φ_p, which is formally defined next.

Definition 4.7   (Spherical contact distribution function.) For a convex compact set A in $\mathbb{R}^2$ that contains the origin and $\nu_2(A) > 0$, the contact distribution function $H_A(r)$ of a stationary point process Φ is given by

$$H_A(r) = 1 - \mathbb{P}(N_p(rA) = 0). \tag{4.12}$$

If $A = B(o, 1)$, then this distribution is called the spherical contact distribution function or the empty space function and is denoted by $H_s(r)$. This is the CDF of the contact distance.

The spherical contact distribution function $H_s(r)$ of the PLCP Φ_p is derived in the following lemma.

Lemma 4.8   *The spherical contact distribution function of the PLCP Φ_p is*

$$H_s(r) = 1 - \exp\left[-2\pi\lambda_l \int_0^r \left(1 - \exp(-2\lambda_p\sqrt{r^2 - \rho^2})\right)\mathrm{d}\rho\right]. \tag{4.13}$$

Proof. By definition, the spherical contact distribution function can be computed as

$$H_s(r) = 1 - \mathbb{P}(N_p(B(o, r)) = 0).$$

So, we now have to compute the probability that there are no points of Φ_p inside the disc $B(o, r)$, as shown in Fig. 4.2. Using the void probability of the PLCP given in Lemma 4.1, we have

$$H_s(r) = 1 - \exp\left[-\lambda_l \int_0^{2\pi} \int_{\mathbb{R}+} \left[1 - \exp\left(-\lambda_p \nu_1 \left(L_{(\rho,\theta)} \cap B(o, r)\right)\right)\right]\mathrm{d}\rho\mathrm{d}\theta\right].$$

For a line L at a distance $\rho \leq r$ from the origin, the length of the line segment inside the disc $B(o, r)$ is $2\sqrt{r^2 - \rho^2}$. Thus, we obtain the spherical contact distribution as

$$H_s(r) = 1 - \exp\left[-\lambda_l \int_0^{2\pi} \int_0^r \left[1 - \exp\left(-2\lambda_p\sqrt{r^2 - \rho^2}\right)\right]\mathrm{d}\rho\mathrm{d}\theta\right].$$

As the integrand in the above equation is independent of θ, the expression for the spherical contact distribution function can be simplified as

$$H_s(r) = 1 - \exp\left[-2\pi\lambda_l \int_0^r \left(1 - \exp(-2\lambda_p\sqrt{r^2 - \rho^2})\right)\mathrm{d}\rho\right]. \tag{4.14}$$

$\square$

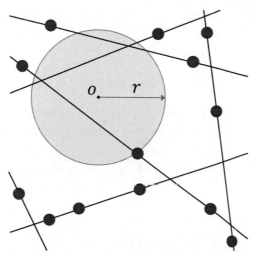

Figure 4.2: Illustration of the spherical contact distance of the PLCP.

4.2.7   NEAREST-NEIGHBOR DISTANCE DISTRIBUTION

The nearest-neighbor distance R_{nn} is the Euclidean distance between a reference point $x \in \Phi_p$ and its closest neighbor in Φ_p, i.e., $R_{nn} = \min_{y \in \Phi_p \setminus \{x\}}\{\|y - x\|\}$. As the PLCP Φ_p is stationary, it is sufficient to condition on a point being at the origin o. The nearest-neighbor distance distribution is often useful in studying connectivity in wireless networks (as will be evident in the following chapters). The CDF of R_{nn} is presented in the following lemma.

Lemma 4.9   *The CDF of the nearest-neighbor distance of Φ_p is*

$$F_{R_{nn}}(r) = 1 - \exp\left[-2\lambda_p r - 2\pi\lambda_l \int_0^r \left(1 - \exp(-2\lambda_p \sqrt{r^2 - \rho^2})\right)d\rho\right]. \qquad (4.15)$$

Proof. By definition, we have

$$F_{R_{nn}}(r) = 1 - \mathbb{P}(R_{nn} > r) = 1 - \mathbb{P}\left(\min_{y \in \Phi_p \setminus \{o\}}\{\|y\|\} > r\right).$$

We now need to compute the probability of the event that the nearest neighbor to the origin is farther than r. In other words, we have to find the probability that there are no points inside the disc $B(o, r)$ under the reduced Palm distribution of the PLCP. Thus, the CDF of R_{nn} can be computed as

$$F_{R_{nn}}(r) = 1 - \mathbb{P}\left(N_p(B(o, r) \cap \Phi_{po}^!) = 0\right). \qquad (4.16)$$

Recall that the distribution of points under the reduced Palm probability $\Phi_{p_0}^!$ is the superposition of the PLCP Φ_p and an independent 1D PPP Ψ_{L_0}. Owing to the independent distribution of points on the line L_0, the above equation can be written as

$$F_{R_{nn}}(r) = 1 - \mathbb{P}\left(N_p(B(o,r) \cap \Psi_{L_0}) = 0\right) \mathbb{P}\left(N_p(B(o,r) \cap \Phi_p) = 0\right).$$

Using the expressions for the void probability of the 1D PPP Ψ_{L_0} and the void probability of the PLCP Φ_p, we obtain the CDF of the nearest-neighbor distance as

$$F_{R_{nn}}(r) = 1 - \exp\left[-2\lambda_p r - 2\pi\lambda_l \int_0^r \left(1 - \exp(-2\lambda_p \sqrt{r^2 - \rho^2})\right)d\rho\right]. \qquad (4.17)$$

$\square$

4.3 LAPLACE FUNCTIONAL

In this section, we will present the Laplace functional of the PLCP Φ_p, which is a random measure analogue of Laplace transform and this result is quite useful in the characterization of aggregate interference power in wireless networks. This result has also been provided in [41, 64].

Lemma 4.10 *Let f be a bounded non-negative measurable function. The Laplace functional of the PLCP Φ_p is*

$$\mathcal{L}(f) = \exp\left[-\lambda_l \int_0^{2\pi} \int_{\mathbb{R}^+} 1 - \exp\left(-\lambda_p \int_{\mathbb{R}} 1 - e^{-g(\rho,\theta,u)} du\right) d\rho d\theta\right], \qquad (4.18)$$

where $g(\rho, \theta, u) = f(\rho\cos\theta + u\sin\theta, \rho\sin\theta - u\cos\theta)$.

Proof. By definition, the Laplace functional of the PLCP is given by

$$\mathcal{L}(f) \triangleq \mathbb{E}_{\Phi_p}\left[\exp\left(-\sum_{x \in \Phi_p} f(x)\right)\right]. \qquad (4.19)$$

By conditioning on the line process and expressing the expectation of product of independent terms as product of expectations, the above equation can be written as

$$\mathcal{L}(f) = \mathbb{E}_{\Phi_l}\left[\prod_{L \in \Phi_l} \mathbb{E}_{\Psi_L}\left[\prod_{x \in \Psi_L} e^{-f(x)} \,\bigg|\, \Phi_l\right]\right]. \qquad (4.20)$$

Note that the location of points of the PLCP can also be characterized by the line on which they are located and their location on the line w.r.t. the foot of the perpendicular onto the line from the origin. For instance, the point at $x \equiv (x, y)$ can be referred to as the point located on

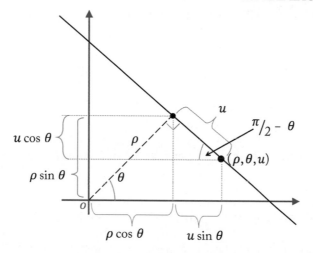

Figure 4.3: Illustration of an alternative representation of the points of the PLCP.

the line L, with parameters (ρ, θ), at a signed distance u from the foot of the perpendicular, as depicted in Fig. 4.3. Thus, the coordinates of a point can also be expressed in terms of ρ, θ, and u. The relation between the Cartesian coordinates and the polar coordinates is given by

$$x = \rho \cos \theta + u \sin \theta,$$
$$y = \rho \sin \theta - u \cos \theta.$$

Upon expressing the location of the points in polar coordinates and using the PGFL of the 1D PPP on each line L, the expression for the Laplace functional in (4.20) simplifies to

$$\mathcal{L}(f) = \mathbb{E}_{\Phi_l}\left[\prod_{(\rho,\theta)\in\Phi_C} \exp\left(-\lambda_p \int_{\mathbb{R}} 1 - e^{-g(\rho,\theta,u)} \mathrm{d}u \right) \right].$$

Using the PGFL of the 2D PPP Φ_C corresponding to the line process Φ_l, the Laplace functional of the PLCP is obtained as

$$\mathcal{L}(f) = \exp\left[-\lambda_l \int_0^{2\pi} \int_{\mathbb{R}+} 1 - \exp\left(-\lambda_p \int_{\mathbb{R}} 1 - e^{-g(\rho,\theta,u)} \mathrm{d}u \right) \mathrm{d}\rho\mathrm{d}\theta \right].$$

$\square$

We can specialize the result of Lemma 4.10 to radially symmetric functions as given in the following Corollary.

Corollary 4.11   If f is a radially symmetric function, then the Laplace functional of Φ_p is given by

$$\mathcal{L}(f) = \exp\left[- 2\pi\lambda_l \int_{\mathbb{R}+} 1 - \exp\left(-\lambda_p \int_{\mathbb{R}} 1 - e^{-g(\sqrt{\rho^2+u^2})} du \right) d\rho \right], \qquad (4.21)$$

where $g(\sqrt{\rho^2 + u^2}) = f(\sqrt{x^2 + y^2})$.

Proof. The proof simply follows from the fact that the functions f and g are independent of θ. $\qquad\square$

Using the result presented in Lemma 4.10, we can also obtain the Laplace functional of the PLCP Φ_p under the reduced Palm distribution as given in the following Lemma.

Lemma 4.12   *Under the reduced Palm distribution, the Laplace functional of the PLCP Φ_p is*

$$\hat{\mathcal{L}}(f) = \exp\left[-\lambda_p \int_{\mathbb{R}} 1 - e^{-g(0,0,u)} du \right.$$

$$\left. - \lambda_l \int_0^{2\pi} \int_{\mathbb{R}+} 1 - \exp\left(-\lambda_p \int_{\mathbb{R}} 1 - e^{-g(\rho,\theta,u)} du \right) d\rho d\theta \right], \quad (4.22)$$

where $g(\rho, \theta, u) = f(\rho\cos\theta + u\sin\theta, \rho\sin\theta - u\cos\theta)$.

Proof. In addition to the steps followed in the proof of Lemma 4.10, we need to handle the independent 1D PPP on the line L_0. Further, without loss of generality, we can consider that L_0 is aligned along the vertical axis (i.e., $\theta_0 = 0$) due to the isotropy of Φ_l. $\qquad\square$

4.4   ASYMPTOTIC CHARACTERISTICS

In this section, we will study the asymptotic behavior of the PLCP for extreme values of line and point densities. This discussion will be useful in developing accurate approximations of this model that could aid in its tractable analysis for some applications. We will explore the asymptotic behavior of the PLCP for both high and low line densities using Choquet's theorem [4]. Choquet's theorem states that the distribution of a random closed set is completely determined by the capacity functional, which is defined as the complement of the void probability. Therefore, we will analyze the asymptotic properties of the PLCP by examining the void probability for extreme values of line and point densities. We will now show that the PLCP asymptotically converges to a 2D PPP in the following theorem.

Theorem 4.13   As the line density approaches infinity ($\lambda_l \to \infty$) and point density tends to zero ($\lambda_p \to 0$) while the overall average number of points per unit area remains unchanged, the

PLCP under reduced Palm distribution converges to a homogeneous 2D PPP with the same point density $\pi \lambda_l \lambda_p$.

Proof. As mentioned earlier, by Choquet's theorem, we know that the distribution of points in a point process is completely characterized by its void probability. Therefore, in order to prove this result, it is sufficient to show that the void probability of the PLCP converges to that of a 2D PPP. So, we will now apply the limits $\lambda_l \to \infty$ and $\lambda_p \to 0$ on the expression for void probability under the reduced Palm distribution given in Corollary 4.2. As the overall density of points $\lambda_a = \pi \lambda_l \lambda_p$ remains constant, the application of the two limits $\lambda_l \to \infty$ and $\lambda_p \to 0$ can be simplified to a single limit by substituting $\lambda_p = \frac{\lambda_a}{\pi \lambda_l}$ in the expression for void probability in (4.5). Thus, the asymptotic void probability can be evaluated as

$$
\lim_{\lambda_l \to \infty} \mathbb{P}\big(N_p(A) = 0\big) = \lim_{\lambda_l \to \infty} \exp\Bigg[-\lambda_p \nu_1(L_0 \cap A) - \lambda_l
$$
$$
\times \int_0^{2\pi} \int_{\mathbb{R}^+} \big[1 - \exp\left(-\lambda_p \nu_1\left(L_{(\rho,\theta)} \cap A\right)\right)\big] \mathrm{d}\rho\mathrm{d}\theta \Bigg]
$$
$$
= \exp\Bigg[\lim_{\lambda_l \to \infty} -\frac{\lambda_a}{\pi \lambda_l} \nu_1(L_0 \cap A) + \lim_{\lambda_l \to \infty} -\lambda_l
$$
$$
\times \int_0^{2\pi} \int_{\mathbb{R}^+} \big[1 - \exp\left(-\frac{\lambda_a}{\pi \lambda_l} \nu_1\left(L_{(\rho,\theta)} \cap A\right)\right)\big] \mathrm{d}\rho\mathrm{d}\theta \Bigg].
$$

Note that the first limit in the above expression evaluates to 0. Further, using the Taylor series expansion of the exponential function, the above equation can be written as

$$
\lim_{\lambda_l \to \infty} \mathbb{P}\big(N_p(A) = 0\big) = \exp\Bigg[\lim_{\lambda_l \to \infty} \lambda_l \int_0^{2\pi} \int_{\mathbb{R}^+} \sum_{k=1}^{\infty} \frac{\left(-\lambda_a \nu_1\left(L_{(\rho,\theta)} \cap A\right)\right)^k}{(\pi \lambda_l)^k \, k!} \mathrm{d}\rho\mathrm{d}\theta \Bigg].
$$

We will handle the terms corresponding to $\{k = 1\}$ and $\{k \geq 2\}$ appearing in the summation in the integrand in the above equation, separately. Hence, we obtain

$$
\lim_{\lambda_l \to \infty} \mathbb{P}\big(N_p(A) = 0\big) = \exp\Bigg[\lim_{\lambda_l \to \infty} \lambda_l \int_0^{2\pi} \int_{\mathbb{R}^+} \frac{\left(-\lambda_a \nu_1\left(L_{(\rho,\theta)} \cap A\right)\right)}{\pi \lambda_l} \mathrm{d}\rho\mathrm{d}\theta
$$
$$
+ \sum_{k=2}^{\infty} \lim_{\lambda_l \to \infty} \int_0^{2\pi} \int_{\mathbb{R}^+} \frac{\left(-\lambda_a \nu_1\left(L_{(\rho,\theta)} \cap A\right)\right)^k}{\pi^k \lambda_l^{k-1} k!} \mathrm{d}\rho\mathrm{d}\theta \Bigg].
$$

We will now focus on the second term in the RHS of the above equation, which can be simplified by applying the Dominated Convergence Theorem (DCT). According to the DCT, if a sequence of real-valued measurable functions $\{f_n\}$ converges pointwise to a function f and is dominated by some integrable function in the sense that $|f_n(x)| \leq g(x)$, $\forall n$ in the index set and $\forall x \in S$, then $\lim_{n \to \infty} \int_S f_n \mathrm{d}\mu = \int_S f \mathrm{d}\mu$. Now, let us examine the integrand in the second term in the previous equation. We can clearly see that the integrand converges to 0 as $\lambda_l \to \infty$ for all $k \geq 2$. Also, as the integrand evaluates to a finite value for all λ_l, we can say that it is dominated by some integrable function. Hence, by applying the DCT, we obtain the asymptotic void probability as

$$\lim_{\lambda_l \to \infty} \mathbb{P}\left(N_p(A) = 0\right) = \exp\left[\lim_{\lambda_l \to \infty} \lambda_l \int_0^{2\pi} \int_{\mathbb{R}^+} \frac{\left(-\lambda_a \nu_1\left(L_{(\rho,\theta)} \cap A\right)\right)}{\pi \lambda_l} d\rho d\theta \right.$$
$$\left. + \sum_{k=2}^{\infty} \int_0^{2\pi} \int_{\mathbb{R}^+} \lim_{\lambda_l \to \infty} \frac{\left(-\lambda_a \nu_1\left(L_{(\rho,\theta)} \cap A\right)\right)^k}{\pi^k \lambda_l^{k-1} k!} d\rho d\theta \right].$$

As discussed, upon applying the limit, the integrand evaluates to 0 for all $k \geq 2$. Thus, the above equation simplifies to

$$\lim_{\lambda_l \to \infty} \mathbb{P}\left(N_p(A) = 0\right) = \exp\left[\lim_{\lambda_l \to \infty} \frac{-\lambda_a}{\pi \lambda_l} \int_0^{2\pi} \int_{\mathbb{R}^+} \left(\nu_1\left(L_{(\rho,\theta)} \cap A\right)\right) \lambda_l d\rho d\theta\right].$$

By applying Campbell's theorem for sum over stationary PPP Φ_C in $\mathcal{C}$, the integral in the above equation can be expressed as the following expectation:

$$\lim_{\lambda_l \to \infty} \mathbb{P}\left(N_p(A) = 0\right) = \exp\left[\lim_{\lambda_l \to \infty} -\frac{\lambda_a}{\pi \lambda_l} \mathbb{E}\left[\sum_{(\rho,\theta) \in \Phi_C} \nu_1\left(L_{(\rho,\theta)} \cap A\right)\right]\right].$$

The expectation in the above expression denotes the average total length of line segments inside the planar set A. Recall that the line density of a line process is nothing but the mean line length per unit area. Thus, the above expectation is simply the product of line density and the area of A. Thus, we have

$$\lim_{\lambda_l \to \infty} \mathbb{P}\left(N_p(A) = 0\right) = \exp\left[\lim_{\lambda_l \to \infty} -\frac{\lambda_a}{\pi \lambda_l}\left(\pi \lambda_l \nu_2(A)\right)\right] = \exp(-\lambda_a \nu_2(A)).$$

The final expression is nothing but the void probability of a homogeneous 2D PPP with density $\lambda_a = \pi \lambda_l \lambda_p$. Thus, we showed that the PLCP asymptotically converges to a 2D PPP. $\square$

This result is pictorially demonstrated in Fig. 4.4.

Remark 4.14 In Theorem 4.13, we showed the convergence of the PLCP to a 2D PPP under the reduced Palm distribution in order to have the same setting as in Theorem 4.15. Note that this result also holds for the regular (non-Palm) distribution of the PLCP.

We will now show the convergence of the PLCP under reduced Palm distribution to a 1D PPP in the following theorem.

Theorem 4.15 As the line density tends to zero ($\lambda_l \to 0$), the PLCP under reduced Palm distribution converges to a homogeneous 1D PPP with density λ_p.

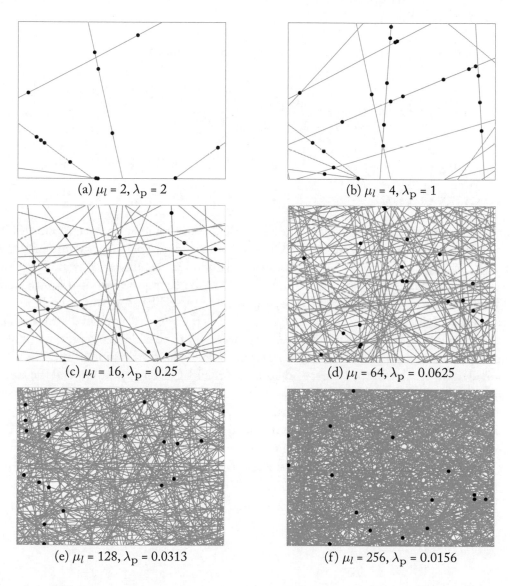

(a) $\mu_l = 2$, $\lambda_p = 2$

(b) $\mu_l = 4$, $\lambda_p = 1$

(c) $\mu_l = 16$, $\lambda_p = 0.25$

(d) $\mu_l = 64$, $\lambda_p = 0.0625$

(e) $\mu_l = 128$, $\lambda_p = 0.0313$

(f) $\mu_l = 256$, $\lambda_p = 0.0156$

Figure 4.4: Illustration of the convergence of the PLCP to a 2D PPP as the line density increases and the point density decreases while the average number of points per unit area remains unchanged.

Proof. Following the same approach as in Theorem 4.13, we can prove this result by applying the limit $\lambda_l \to 0$ to the expression for void probability of the PLCP under the reduced Palm distribution. Thus, we have

$$\lim_{\lambda_l \to 0} \mathbb{P}(N_p(A) = 0) = \exp\left[\lim_{\lambda_l \to 0} -\lambda_p \nu_1(L_0 \cap A) - \lim_{\lambda_l \to 0} \lambda_l \right.$$
$$\left. \times \int_0^{2\pi} \int_{\mathbb{R}^+} \left[1 - \exp\left(-\lambda_p \nu_1\left(L_{(\rho,\theta)} \cap A\right)\right) \right] d\rho d\theta \right]$$
$$= \exp\left[-\lambda_p \nu_1(L_0 \cap A) \right],$$

which is nothing but the expression of the void probability of a 1D PPP with density λ_p. $\square$

4.5   SUMMARY

In this chapter, we studied the construction of the PLCP and some of its basic properties. We discussed the motion-invariance of the PLCP and its dependence on the structure of the underlying PLP. We then derived the expression for the void probability of a motion-invariant PLCP, which completely describes the point process. We then computed the PMF of the number of points of the PLCP in a disc of fixed radius and the typical Poisson Voronoi cell. We also discuss key distributional properties of the PLCP, such as the spherical contact distribution and the nearest-neighbor distance distribution. As it is necessary to consider the typical point of the point process in several applications, we also presented the Palm distribution of the PLCP. We then provided the results for the Laplace functional of the PLCP. Finally, we studied the asymptotic behavior of the PLCP, where we showed the convergence of the PLCP to homogeneous 1D and 2D PPPs under different conditions.

Having discussed the construction and several fundamental properties of the PLCP, we will now proceed to explore useful applications of this spatial model. We begin with the application of PLCP to vehicular communication networks in the next chapter.

CHAPTER 5

Vehicular Communication Networks

In Chapters 2–4, we discussed the theory of PPP and PLP, which are the building blocks of the PLCP and provided a rigorous exposition of the theory of the PLCP, which constitute the first part of this book. From this chapter onward, which constitutes the second part of the book, we will focus on the applications of the PLCP to vehicular networks. In this chapter, we present a general setting of vehicular communication network, which will be used to analyze key performance metrics for a variety of network settings in the subsequent chapters. We recall the various components and modes of communication in a vehicular network next.

A vehicular communication network comprises vehicular nodes, RSUs, and MBSs that communicate with each other to share information that could be useful in improving road safety and assist in intelligent navigation, as illustrated in Fig. 5.1 [79–81]. For instance, information about an accident can be communicated to other vehicles traveling on the same road to avoid rear end collisions. Also, the prior knowledge of such events could help the road users to adjust their travel routes in advance, thereby avoiding traffic congestions. Such applications are enabled by various modes of communication in a vehicular network such as vehicle-to-vehicle (V2V), vehicle-to-infrastructure (V2I), vehicle-to-pedestrian (V2P), and vehicle-to-network (V2N) communication [82, 83]. Vehicular communication networks are distinguished by their stringent requirements of high reliability and low latency because of which it is necessary to carefully evaluate their performance in different operational regimes [35].

As discussed in Chapter 1, vehicular networks have a peculiar spatial geometry as the locations of vehicular nodes and RSUs are coupled with the layout of the underlying road network. Hence, the PLCP is a well-suited model for vehicular networks, where the lines of the PLP model the roads and the points represent the locations of the vehicular nodes and RSUs. Using the PLCP model, the basic performance metrics, such as coverage and rate, have been analyzed in [35–41, 68, 69, 96]. We will discuss some of the key results from these recent works in this book.

5.1 OBJECTIVES

In this book, we will mainly focus on the characterization of the success probability of the typical communication link or the coverage probability of the typical receiver in a vehicular network. The communication between a transmitter and a receiver node is declared a success (or equiv-

Figure 5.1: Illustration of various components of a vehicular communication network.

alently the receiver is said to be in coverage) if the SINR measured at the receiver exceeds a predetermined threshold. We will study the SINR-based performance of a vehicular communication network for the ad hoc and cellular models in the next two chapters where the locations of vehicular nodes and RSUs are modeled using a PLCP. The doubly stochastic nature of the PLCP introduces additional complexity into the coverage analysis of the network as compared to the canonical setup where the locations of wireless nodes are modeled by PPPs. Having said that, it is important to note that the basic analytical framework for the coverage analysis largely remains the same as that of the canonical models. Therefore, before we delve into the intricacies and technical challenges posed by the PLCP model in the next two chapters, it will be useful to provide an overview of the analytical procedure to compute the coverage probability, which is the main purpose of this chapter. In order to explain this concretely, we formulate the coverage probability problem for a general system model and identify two key results that would be required in the coverage analysis of the PLCP settings studied in the next two chapters. Some of the key modeling assumptions and notation are presented in the next section.

5.2   NOTATION AND MODELING ASSUMPTIONS

Let us denote the set of locations of all the transmitter and receiver nodes in the network by stationary point processes $\Phi_T \equiv \{w\}$ and $\Phi_R \equiv \{z\}$, respectively. While we are eventually interested in the computations where Φ_T and Φ_R are both PLCPs driven by the same underlying

PLP, we do not need to make this assumption in the current chapter. Now, owing to the stationarity of this setting, the typical receiver node can be placed at the origin o. For this setting, the SINR observed at the typical receiver can be expressed as:

$$\text{SINR} = \frac{P_{w^*} H_{w^*} R^{-\alpha}}{\sum_{w \in \Phi_T \setminus w^*} P_w H_w \|w\|^{-\alpha} + \sigma^2}, \tag{5.1}$$

where the numerator represents the received signal power from the desired transmitter, which is often referred to as the *serving node*. The first term in the denominator represents the aggregate interference power measured at the receiver, and the second term in the denominator represents the noise power σ^2. We will now define all the random variables appearing in (5.1). P_{w^*} and P_w denote the transmit power of the serving node located at $w^* \in \Phi_T$ and interfering transmitter located at w, respectively. We assume a single-slope standard power-law path-loss model with exponent α. The channel fading gains for the links from the typical receiver to the desired transmitter and interfering transmitters are denoted by H_{w^*} and H_w, respectively. In the interest of expositional simplicity, we assume that all the wireless links suffer from Rayleigh fading and the fading gains are exponentially distributed with unit mean. The distance between the typical receiver and the serving node, referred to as the serving distance, is denoted by $R = \|w^*\|$.

5.3 COVERAGE PROBABILITY

As mentioned earlier, our objective is to compute the coverage probability of the typical receiver. For the setup introduced in the previous section, the coverage probability can be computed as

$$P_c = \mathbb{P}(\text{SINR} > \beta) = \mathbb{P}\left(\frac{P_{w^*} H_{w^*} R^{-\alpha}}{I + \sigma^2} > \beta\right), \tag{5.2}$$

where β denotes the SINR threshold and I denotes the aggregate interference at the typical receiver. By conditioning on the serving distance R and the interference I, the coverage probability can be written as

$$P_c = \mathbb{E}_R \mathbb{E}_I \left[\mathbb{P}\left(H_{w^*} > \frac{\beta r^\alpha}{P_{w^*}}(I + \sigma^2) \mid R, I \right) \right]. \tag{5.3}$$

From the CCDF of the exponential random variable H_{w^*}, it follows that

$$P_c = \mathbb{E}_R \mathbb{E}_I \left[e^{-\frac{\beta r^\alpha}{P_{w^*}} I} e^{-\frac{\beta r^\alpha}{P_{w^*}} \sigma^2} \right] = \mathbb{E}_R \left[e^{-\frac{\beta r^\alpha}{P_{w^*}} \sigma^2} \mathbb{E}_I \left[e^{-\frac{\beta r^\alpha}{P_{w^*}} I} \right] \right]. \tag{5.4}$$

The expectation w.r.t. I in the above equation can be expressed in terms of the Laplace transform of the interference power distribution. Thus, the coverage probability of the typical receiver is given by

$$P_c = \int_{\mathbb{R}} e^{-\frac{\beta r^\alpha}{P_{w^*}} \sigma^2} \mathcal{L}_I\left(\frac{\beta r^\alpha}{P_{w^*}}\right) f_R(r) \mathrm{d}r. \tag{5.5}$$

Thus, the two key pieces of information necessary to compute the coverage probability are: (i) the serving distance distribution, and (ii) the Laplace transform of interference power distribution conditioned on the serving distance. Note that this expression of coverage probability is agnostic to the spatial model. However, the distributions of the serving distance and the interference power depend on the spatial distribution of nodes and association policies and hence, they would vary across different models. In this book, we will focus on the computation of the coverage probability for two types of network models namely, (i) ad hoc network model and (ii) cellular network model in Chapters 6 and 7, respectively.

5.4 SUMMARY

In this chapter, we gave a brief introduction to vehicular communication networks and presented the motivation behind the spatial modeling of vehicular networks using PLCP. In this book, our main focus is on the computation of SINR-based coverage probability of the typical receiver in the network. Toward this goal, we introduced some of the modeling assumptions pertaining to the signal propagation considered in our analysis. We then provided a general expression for coverage probability that is applicable to any stationary point processes of transmitters and receivers. The two key pieces of information that are necessary to compute the coverage probability are the distribution of the serving distance and the Laplace transform of the distribution of interference power conditioned on the serving distance. Building on these insights, we will derive the coverage probability for both ad hoc and cellular network models in the next two chapters.

CHAPTER 6

Ad Hoc Network Model

In a VANET, the vehicular nodes communicate with each other without any support from the network infrastructure. A key quality of service (QoS) metric that is often used in characterizing the performance of ad hoc networks is the success probability of the typical link, which is defined as the probability with which the SINR measured at the receiver of the typical link exceeds a predetermined threshold. In this chapter, we first present this result for the canonical Poisson bipolar model, which has been the most commonly used spatial model in the analysis of ad hoc networks. We then derive the success probability of the typical link in a VANET modeled using the PLCP. This analysis reveals the relation between the network performance and the key system parameters, such as the densities of roads and nodes.

6.1   POISSON BIPOLAR MODEL

We model the locations of the wireless nodes by a homogeneous 2D PPP Ψ with density λ_n. Assuming that each node transmits with a probability p independent of the other nodes in the network, the locations of transmitting nodes form a thinned PPP Ψ_t with density $\lambda_v = p\lambda_n$. We further assume that the receiver nodes are located at a fixed distance d from their respective transmitters in arbitrary directions, thereby forming a Poisson bipolar network, as depicted in Fig. 6.1. The receiver node of the typical link in this network is referred to as the typical receiver. Without loss of generality, we assume that the typical receiver is located at the origin. Assuming that all the nodes have the same transmit power P_t, the SINR at the typical receiver is given by

$$\text{SINR} = \frac{P_t H_0 d^{-\alpha}}{\sum_{w \in \Psi_t} P_t H_w \|w\|^{-\alpha} + \sigma^2}. \tag{6.1}$$

Since the serving distance is fixed, we only need to calculate the Laplace transform of interference power distribution to compute the success probability of the typical link. The Laplace transform of interference power distribution for this setup is well known in the literature [6] and is given in the following lemma.

Lemma 6.1   *The Laplace transform of the interference power distribution is*

$$\mathcal{L}_I(s) = \exp\left[-\pi p\lambda_n (sP_t)^{\frac{2}{\alpha}} \frac{2\pi}{\alpha} \csc\left(\frac{2\pi}{\alpha}\right)\right]. \tag{6.2}$$

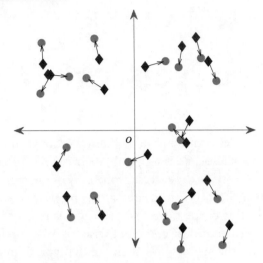

Figure 6.1: Illustration of the Poisson bipolar model.

Proof. The Laplace transform of the distribution of the interference power can be computed as

$$\mathcal{L}_I(s) = \mathbb{E}_I\left[e^{-sI}\right] = \mathbb{E}_{\Psi_t}\mathbb{E}_{H_w}\left[\exp\left(-s\sum_{w\in\Psi_t} P_t H_w \|w\|^{-\alpha}\right)\right].$$

First, computing the expectation w.r.t. H_w, we obtain

$$\mathcal{L}_I(s) = \mathbb{E}_{\Psi_t}\left[\prod_{w\in\Psi_t}\frac{1}{1+sP_t\|w\|^{-\alpha}}\right].$$

Using the PGFL of the homogeneous 2D PPP Ψ_t, the Laplace transform of interference power distribution is then obtained as

$$\mathcal{L}_I(s) = \exp\left(-2\pi\lambda_v\int_0^\infty \frac{sP_t w^{1-\alpha}}{1+sP_t w^{-\alpha}}\mathrm{d}w\right).$$

Solving the integral in the above equation, we obtain the expression for the Laplace transform of the interference power distribution. □

Substituting (6.2) in (5.5), the closed-form expression for success probability of the typical link is obtained as

$$P_c = \exp\left[-\frac{\beta d^\alpha}{P_t}\sigma^2 - \pi p\lambda_n\beta^{\frac{2}{\alpha}}d^2\frac{2\pi}{\alpha}\csc\left(\frac{2\pi}{\alpha}\right)\right]. \tag{6.3}$$

Despite the relevance of the Poisson bipolar model in the analysis of ad hoc networks in general, it is not suitable for VANETs. That said, it may serve as a useful approximation under certain conditions, as will be discussed later in this chapter.

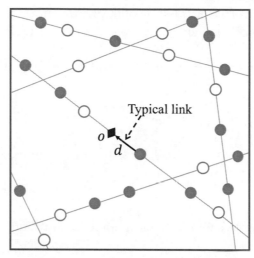

Figure 6.2: Illustration of the system model of a VANET modeled by a Poisson line Cox bipolar model. The filled and hollow circles represent active and inactive transmitters in the network, respectively.

6.2   POISSON LINE COX BIPOLAR (PLCB) MODEL

We will now present the analysis of success probability of the typical link in a VANET using PLCB model. We model the locations of the vehicular nodes by a motion-invariant PLCP, where the spatial layout of roads is modeled by PLP Φ_l with line density μ_l and the location of the nodes on each road is modeled by a homogeneous 1D PPP with density λ_n. Assuming that each node transmits independently with a probability p, the locations of the transmitting nodes on each line L form a thinned PPP Ψ_L with density $p\lambda_n$. Further, we assume that the receiver nodes are located at a fixed distance d from their respective transmitters on the same line, thereby forming a PLCB network, as depicted in Fig. 6.2. We assume that the receiver node of the typical link is located at the origin. Thus, under the reduced Palm distribution of the PLCP, the spatial distribution of the transmitting nodes is given by $\Phi_{t_0} \equiv \cup_{L \in \Phi_{l_0}} \Psi_L$, where $\Phi_{l_0} = \Phi_l \cup \{L_0\}$. We assume that all the nodes have the same transmit power P_t. For this setup, the SINR measured at the receiver node of the typical link is

$$\text{SINR} = \frac{P_t H_0 d^{-\alpha}}{\sum_{L \in \Phi_{l_0}} \sum_{\text{w} \in \Psi_L} P_t H_\text{w} \|\text{w}\|^{-\alpha} + \sigma^2}. \tag{6.4}$$

The Laplace transform of the distribution of the interference at the typical receiver is computed in the following lemma by leveraging the properties of the PLCP presented in Section 4.2.

Lemma 6.2 *The Laplace transform of the distribution of the interference power is*

$$\mathcal{L}_I(s) = \exp\left[-2p\lambda_n \int_0^\infty \frac{1}{1 + (sP_t)^{-1}u^\alpha} du \right.$$
$$\left. - 2\mu_l \int_0^\infty 1 - \exp\left[-\int_0^\infty \frac{2p\lambda_n P_t s}{P_t s + (\rho^2 + u^2)^{\alpha/2}} du \right] d\rho \right]. \quad (6.5)$$

Proof. By definition, the Laplace transform of interference distribution is computed as

$$\mathcal{L}_I(s) = \mathbb{E}\left[\exp\left(-s \sum_{w \in \Phi_{t_0}} P_t H_w \|w\|^{-\alpha} \right) \right]$$
$$= \mathbb{E}_{\Phi_{t_0}} \mathbb{E}_{H_w}\left[\prod_{w \in \Phi_{t_0}} \exp\left(-s P_t H_w \|w\|^{-\alpha} \right) \right]. \quad (6.6)$$

By taking the expectation w.r.t. H_w first, we obtain

$$\mathcal{L}_I(s) = \mathbb{E}_{\Phi_{t_0}}\left[\prod_{w \in \Phi_{t_0}} \frac{1}{1 + s P_t \|w\|^{-\alpha}} \right]. \quad (6.7)$$

Note that the above expression can be easily simplified by using the Laplace functional of the PLCP under reduced Palm distribution, which is given in Lemma 4.12. First, we express the Cartesian coordinates w in terms of the parameters (ρ, θ, u) as explained in Section 4.3. So, the Euclidean distances of the points of the PLCP from the origin in the above expression can be written as $\|w\| = \sqrt{\rho^2 + u^2}$. Thus, using (4.22), the Laplace transform of the interference power distribution is obtained as

$$\mathcal{L}_I(s) = \exp\left[-2p\lambda_n \int_0^\infty \frac{1}{1 + (sP_t)^{-1}u^\alpha} du \right.$$
$$\left. - 2\mu_l \int_0^\infty 1 - \exp\left[-\int_0^\infty \frac{2p\lambda_n P_t s}{P_t s + (\rho^2 + u^2)^{\alpha/2}} du \right] d\rho \right]. \quad (6.8)$$

This completes the proof. □

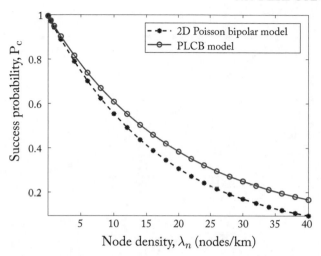

Figure 6.3: Success probability of the typical link as a function of node density ($\mu_l = 30$ km^{-1}, $d = 20$ m, $p = 0.5$, and $\alpha = 4$).

Substituting (6.5) in (5.5), the success probability of the typical link in a VANET is obtained as

$$P_c = \exp\left[-\frac{\beta\sigma^2 d^\alpha}{P_t} - 2p\lambda_n \int_0^\infty \frac{\beta d^\alpha u^{-\alpha}}{1 + \beta d^\alpha u^{-\alpha}} du - 2\mu_l \right.$$
$$\left. \times \int_0^\infty 1 - \exp\left[-2p\lambda_n \int_0^\infty \frac{\beta d^\alpha (\rho^2 + u^2)^{-\alpha/2}}{1 + \beta d^\alpha (\rho^2 + u^2)^{-\alpha/2}} du \right] d\rho \right]. \quad (6.9)$$

6.3   PERFORMANCE TRENDS

In this section, we will discuss the trends in the performance of a VANET as a function of the network parameters, such as line density and node density. In Fig. 6.3, we plot the success probability of the typical link in a VANET as a function of node density. As expected, the success probability of the typical link improves as the node density decreases due to the decline in the interference power at the typical receiver. As the node density decreases, we can also observe that the success probability of the typical link converges to that of a 2D Poisson bipolar network with the same average number of transmitter nodes per unit area. This behavior agrees with the results presented in Theorem 4.13, where we showed that the PLCP asymptotically converges to a homogeneous 2D PPP.

We plot the success probability of the typical link as a function of the line density in Fig. 6.4. It can be observed that the success probability of the typical link improves as the line

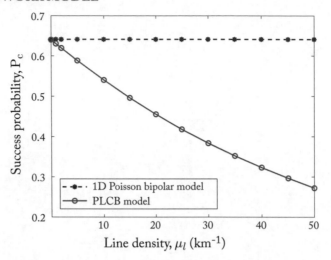

Figure 6.4: Success probability of the typical link as a function of line density ($\lambda_v = 20$ nodes/km, $p = 0.5$, $d = 20$ m, and $\alpha = 4$).

density decreases. This is because of the reduction in interference power at the typical receiver as the distance to interfering vehicular nodes on other lines increases as the line density decreases. This means that the vehicular nodes in areas with dense roads are likely to suffer from poor performance as compared to those in areas with sparse roads for the same density of nodes. Also, as the line density decreases, we observe that the success probability of the typical link in a VANET approaches to that of a 1D Poisson bipolar model with the same vehicular node density. This is because of the fact that the PLCP, under reduced Palm distribution, asymptotically converges to a 1D PPP as discussed in Theorem 4.15.

Area spectral efficiency (ASE). Using the analytical expression for the success probability of the typical link given in (6.9), we can easily compute the ASE of the network, which offers a broader perspective on the overall network performance. The ASE is defined as the average number of successfully transmitted bits per unit time per unit area of the network and can be computed as

$$\text{ASE} = \lambda_{\text{act}} \text{P}_\text{c} \log_2(1 + \beta) \text{ bits/s/Hz/km}^2, \tag{6.10}$$

where $\lambda_{\text{act}} = \mu_l p \lambda_n$ is the average number of active nodes per unit area.

The number of active nodes in the network linearly increases with the transmission probability p. At the same time, the success probability of the typical link in a VANET degrades as the transmission probability increases due to increased interference from the active nodes in the network and this, in turn, deteriorates the ASE. As a result, the conflicting effect of p on the ASE yields an optimum transmission probability p^* that maximizes the ASE, as demonstrated

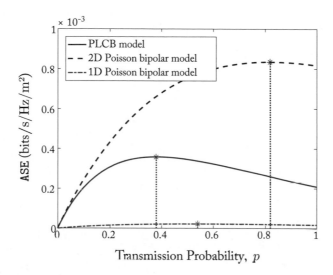

Figure 6.5: ASE of the network as a function of transmission probability p ($\mu_l = 25$ km^{-1}, $\lambda_v = 70$ nodes/km, and $\alpha = 4$).

in Fig. 6.5. Also, it must be noted that the value of p^* and the corresponding ASE obtained from the PLCP model differs significantly from that of conventional 1D and 2D PPP models.

6.4   SUMMARY

In this chapter, we analyzed the performance of a VANET. First, we briefly discussed the performance analysis of the typical link in a canonical setup where an ad hoc network is modeled by a Poisson bipolar model. We then presented the analytical procedure to compute the success probability of the typical link for the PLCB model. It can be observed that the success probability of the typical link in a VANET improves as the line density and node density decreases. We also observed the convergence of the results of PLCB model to that of 1D and 2D Poisson bipolar models for extreme values of line and node densities. Further, we computed the ASE of the network and demonstrated that the optimum transmission probability that maximizes the ASE for the PLCB model is quite different from that of the canonical 1D and 2D PPP models for the same node densities. We now proceed to the cellular network model in the next chapter.

CHAPTER 7

Cellular Network Model

In the previous chapter, we analyzed the performance of the typical link in a VANET where the locations of nodes are modeled by PLCP. In this chapter, we focus on the characterization of SINR-based coverage probability of the typical receiver node for the cellular network model. We consider a multi-tier setup where vehicular nodes and RSUs coexist with cellular MBSs. As is usually the case in cellular networks, we assume that the typical vehicular node would connect to the transmitting node from which it receives the highest average received signal power. The key steps involved in the coverage analysis of a multi-tier network are presented in the flowchart in Fig. 7.1. We first briefly present the canonical coverage analysis of a two-tier conventional cellular network, where the locations of the mobile users, MBSs, and small-cell base stations (SBSs) are modeled by independent and homogeneous 2D PPPs. This discussion on the coverage analysis for the canonical setup will provide the readers an overview of the analytical framework and also help in understanding the additional challenges posed by the PLCP model. We then present the coverage analysis for the vehicular network where the locations of RSUs and vehicular nodes are modeled by PLCP. We also discuss a few design insights based on the trends in the coverage probability as a function of key network parameters.

7.1   2D PPP MODEL

The coverage analysis of a network where the locations of nodes are modeled by 2D PPPs is very well investigated in the literature [14–18] and, hence, we will not delve into the derivation of most of the results for the canonical setup presented in this section.

7.1.1   SYSTEM MODEL

We model the locations of MBSs, SBSs, and mobile users by independent and homogeneous 2D PPPs Φ_1, Φ_2, and Φ_r, with densities λ_1, λ_2, and λ_r, respectively. We will refer to the MBSs and SBSs as tier 1 and tier 2 nodes, respectively. Owing to the stationarity of the homogeneous 2D PPP, we translate the origin to the location of the typical receiver, which is an arbitrarily chosen receiver node from Φ_r. We denote the transmit powers of tier 1 and tier 2 nodes by P_1 and P_2, respectively. We denote the channel fading gains between the typical receiver and the tier 1 and tier 2 nodes by H_1 and H_2, respectively. As mentioned earlier, the typical receiver connects to the node from which it receives the highest average power. Thus, the typical receiver would connect to its closest tier 1 or tier 2 node, located at $x_1^* = \arg\min_{x \in \Phi_1} \|x\|$ and $x_2^* = \arg\min_{x \in \Phi_2} \|x\|$, respectively. Further, we consider selection bias B_1 and B_2 for the tier 1 and

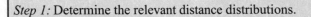

Step 1: Determine the relevant distance distributions.

Step 2: Compute the probability with which the typical receiver connects to i^{th} tier ($i = 1, 2, \dots K$).

Step 3: Determine the distribution of the serving distance conditioned on the event that the typical receiver is connected to the ith tier.

Step 4: Calculate the Laplace transform of interference power distribution conditioned on the serving distance and the event that the typical receiver is connected to the ith tier.

Step 5: Compute coverage probability using law of total probability.

Figure 7.1: Flowchart of steps involved in the computation of coverage probability of the typical receiver for a multi-tier network.

tier 2 nodes, respectively. Thus, the serving node is located at $\mathrm{x}^* = \arg \min_{\mathrm{x} \in \{\mathrm{x}_j^*\}} P_j B_j \|\mathrm{x}\|^{-\alpha}$, where $j \in \{1, 2\}$. We denote the interference from the tier 1 and tier 2 nodes by I_1 and I_2, respectively. For this setup, the SINR measured at the typical receiver is

$$\mathrm{SINR} = \frac{P_r(\mathrm{x}^*)}{I_1 + I_2 + \sigma^2}, \tag{7.1}$$

where

$$P_r(\mathrm{x}^*) = P_{\mathrm{x}^*} H_{\mathrm{x}^*} \|\mathrm{x}^*\|^{-\alpha}, \tag{7.2}$$

$$I_1 = \sum_{\substack{\mathrm{x} \in \Phi_1 \\ \|\mathrm{x}\| > \left(\frac{P_{\mathrm{x}^*} B_{\mathrm{x}^*}}{P_1 B_1}\right)^{-\frac{1}{\alpha}} \|\mathrm{x}^*\|}} P_1 H_1 \|\mathrm{x}\|^{-\alpha}, \tag{7.3}$$

and

$$I_2 = \sum_{\substack{\mathrm{x} \in \Phi_2 \\ \|\mathrm{x}\| > \left(\frac{P_{\mathrm{x}^*} B_{\mathrm{x}^*}}{P_2 B_2}\right)^{-\frac{1}{\alpha}} \|\mathrm{x}^*\|}} P_2 H_2 \|\mathrm{x}\|^{-\alpha}. \tag{7.4}$$

7.1.2   COVERAGE ANALYSIS

Following the steps presented in the flowchart in Fig. 7.1, we will now compute the coverage probability of the typical receiver for this setup.

Step 1: Relevant distance distributions. The key distributions that are of interest to us are the CDF and PDF of the distances between the typical receiver and the candidate serving nodes from tier 1 and tier 2. Let us denote the distance from the typical receiver to x_1^* and x_2^* by R_1 and R_2, respectively. As Φ_1 and Φ_2 are independent homogeneous 2D PPPs, the CDF and PDF of R_i, where $i \in \{1, 2\}$, are given by

$$F_{R_i}(r_i) = 1 - \exp(-\lambda_i \pi r_i^2), \tag{7.5}$$

$$f_{R_i}(r_i) = 2\pi \lambda_i r_i \exp(-\lambda_i \pi r_i^2). \tag{7.6}$$

Step 2: Association probabilities. We denote the events in which the typical receiver connects to the candidate serving node from tier 1 and tier 2 by $\mathcal{E}_1$ and $\mathcal{E}_2$, respectively. The probability of occurrence of the event $\mathcal{E}_1$ can be calculated as

$$\begin{aligned}
\mathbb{P}(\mathcal{E}_1) &= \mathbb{P}\left(P_1 B_1 R_1^{-\alpha} > P_2 B_2 R_2^{-\alpha}\right) \\
&= \int_0^\infty F_{R_1}(\gamma r_2) f_{R_2}(r_2) \mathrm{d}r_2 \\
&= \frac{\lambda_1 \gamma^2}{\lambda_1 \gamma^2 + \lambda_2},
\end{aligned} \tag{7.7}$$

where $\gamma = \left(\frac{P_2 B_2}{P_1 B_1}\right)^{-\frac{1}{\alpha}}$. As the events $\mathcal{E}_1$ and $\mathcal{E}_2$ are complementary, the probability of occurrence of $\mathcal{E}_2$ is $\mathbb{P}(\mathcal{E}_2) = 1 - \mathbb{P}(\mathcal{E}_1)$.

Step 3: Serving distance distribution. In order to characterize the desired signal power, we compute the distribution of the serving distance $R = \|x^*\|$ conditioned on the events $\mathcal{E}_1$ and $\mathcal{E}_2$. The CDF of R conditioned on $\mathcal{E}_1$ can be computed as

$$\begin{aligned}
F_R(r \mid \mathcal{E}_1) &= 1 - \mathbb{P}(R > r \mid \mathcal{E}_1) \\
&= 1 - \frac{1}{\mathbb{P}(\mathcal{E}_1)} \int_{\gamma^{-1}r}^\infty \left[F_{R_1}(\gamma r_2) - F_{R_1}(r)\right] f_{R_2}(r_2) \mathrm{d}r_2 \\
&= 1 - \exp\left(-\lambda_1 \pi r^2 - \lambda_2 \pi \gamma^{-2} r^2\right).
\end{aligned}$$

Thus, the PDF of R conditioned on $\mathcal{E}_1$ is

$$f_R(r \mid \mathcal{E}_1) = 2\pi r(\lambda_1 + \lambda_2 \gamma^{-2}) e^{-\lambda_1 \pi r^2 - \lambda_2 \pi \gamma^{-2} r^2}. \tag{7.8}$$

Similarly, the PDF of R conditioned on $\mathcal{E}_2$ is given by

$$f_R(r \mid \mathcal{E}_2) = 2\pi r(\lambda_1 \gamma^2 + \lambda_2) e^{-\lambda_1 \pi \gamma^2 r^2 - \lambda_2 \pi r^2}. \tag{7.9}$$

Step 4: Laplace transform of interference. In this step, we compute the Laplace transform of the distribution of the aggregate interference power measured at the typical receiver conditioned on the serving distance R and the events $\mathcal{E}_1$ and $\mathcal{E}_2$. As the interference powers due to the tier 1 nodes I_1 and tier 2 nodes I_2 are independent, the Laplace transform of the aggregate interference can be computed as the product of the Laplace transforms of the individual components I_1 and I_2. We will first focus on the Laplace transform of I_1. When the typical receiver is connected to a tier 1 node at a distance R, it means that there are no interfering tier 1 nodes whose distance from the typical receiver is less than R. In other words, there are no tier 1 nodes inside the disc $B(o, R)$. Thus, we have

$$\mathcal{L}_{I_1}(s \mid \mathcal{E}_1) = \mathbb{E}\left[\exp\left(-s \sum_{\mathsf{x} \in \Phi_1 \setminus B(o,r)} P_1 H_1 \|\mathsf{x}\|^{-\alpha}\right)\right]. \tag{7.10}$$

By first taking the expectation w.r.t. H_1 and then using the PGFL of the 2D PPP Φ_1, we obtain

$$\mathcal{L}_{I_1}(s \mid \mathcal{E}_1) = \exp\left(-2\pi\lambda_1 \int_r^\infty \frac{sP_1 x^{1-\alpha}}{1 + sP_1 x^{-\alpha}} dx\right). \tag{7.11}$$

We will now characterize the interference from tier 2 nodes. Conditioning on the event $\mathcal{E}_1$ and serving distance R implies that there cannot be any tier 2 node whose distance from the typical receiver is less than $\gamma^{-1}R$. Therefore, we obtain

$$\mathcal{L}_{I_2}(s \mid \mathcal{E}_1) = \exp\left(-2\pi\lambda_2 \int_{\gamma^{-1}r}^\infty \frac{sP_2 x^{1-\alpha}}{1 + sP_2 x^{-\alpha}} dx\right). \tag{7.12}$$

As discussed earlier, the Laplace transform of the aggregate interference conditioned on $\mathcal{E}_1$ and R can be computed as

$$\begin{aligned}
\mathcal{L}_I(s \mid \mathcal{E}_1) &= \mathcal{L}_{I_1}(s \mid \mathcal{E}_1)\mathcal{L}_{I_2}(s \mid \mathcal{E}_1) \\
&= \exp\left[-2\pi\lambda_1 \int_r^\infty \frac{sP_1 x^{1-\alpha}}{1 + sP_1 x^{-\alpha}} dx - 2\pi\lambda_2 \int_{\gamma^{-1}r}^\infty \frac{sP_2 x^{1-\alpha}}{1 + sP_2 x^{-\alpha}} dx\right].
\end{aligned} \tag{7.13}$$

Similarly, the Laplace transform of the aggregate interference power distribution conditioned on $\mathcal{E}_2$ and R is given by

$$\mathcal{L}_I(s \mid \mathcal{E}_2) = \exp\left[-2\pi\lambda_1 \int_{\gamma r}^\infty \frac{sP_1 x^{1-\alpha}}{1 + sP_1 x^{-\alpha}} dx - 2\pi\lambda_2 \int_r^\infty \frac{sP_2 x^{1-\alpha}}{1 + sP_2 x^{-\alpha}} dx\right]. \tag{7.14}$$

Step 5: Coverage probability. Substituting the above results in (5.5), the coverage probability conditioned on the events $\mathcal{E}_1$ and $\mathcal{E}_2$ are obtained as

$$\mathbb{P}(\mathrm{SINR} > \beta \mid \mathcal{E}_i) = \int_0^\infty e^{-\frac{\beta r^\alpha}{P_i}\sigma^2} \mathcal{L}_I\left(\frac{\beta r^\alpha}{P_i} \mid \mathcal{E}_i\right) f_R(r \mid \mathcal{E}_i) \, dr. \tag{7.15}$$

Using the law of total probability, we obtain the coverage probability as

$$P_c = \mathbb{P}(\mathcal{E}_1)\mathbb{P}(\text{SINR} > \beta \mid \mathcal{E}_1) + \mathbb{P}(\mathcal{E}_2)\mathbb{P}(\text{SINR} > \beta \mid \mathcal{E}_2)$$

$$= \int_0^\infty \exp\left[-\frac{\beta r^\alpha \sigma^2}{P_1} - \lambda_1 \int_r^\infty \frac{2\pi\beta r^\alpha x^{1-\alpha}}{1 + \beta r^\alpha x^{-\alpha}} dx - \lambda_2 \int_{\gamma^{-1}r}^\infty \frac{2\pi\beta r^\alpha P_2 x^{1-\alpha}}{P_1 + \beta r^\alpha P_2 x^{-\alpha}} dx \right]$$

$$\times 2\pi r \lambda_1 e^{-\lambda_1 \pi r^2 - \lambda_2 \pi \gamma^{-2} r^2} dr$$

$$+ \int_0^\infty \exp\left[-\frac{\beta r^\alpha \sigma^2}{P_2} - \lambda_1 \int_{\gamma r}^\infty \frac{2\pi\beta r^\alpha P_1 x^{1-\alpha}}{P_2 + \beta r^\alpha P_1 x^{-\alpha}} dx - \lambda_2 \int_r^\infty \frac{2\pi\beta r^\alpha x^{1-\alpha}}{1 + \beta r^\alpha x^{-\alpha}} dx \right]$$

$$\times 2\pi r \lambda_2 e^{-\lambda_1 \pi \gamma^2 r^2 - \lambda_2 \pi r^2} dr. \tag{7.16}$$

Having presented the coverage analysis of the typical receiver for the canonical setup where the locations of the nodes are modeled by PPPs, we will now discuss the coverage analysis for the PLCP model.

7.2   PLCP MODEL

In this section, we will present the coverage analysis for the typical receiver in a two-tier network where the locations of vehicular nodes and RSUs are modeled by PLCP [35, 40]. In this setup, we consider MBSs as tier 1 nodes and collectively refer to RSUs and transmitting vehicular nodes as tier 2 nodes. Although most of the modeling assumptions are the same as described in the previous section, we still provide a comprehensive description of the system model next for completeness.

7.2.1   SYSTEM MODEL

We model the locations of tier 1 nodes by a homogeneous 2D PPP Φ_1 with density λ_1. We model the locations of tier 2 nodes by a motion-invariant PLCP Φ_v such that the locations of the nodes on each line L of the underlying PLP Φ_l form a homogeneous 1D PPP Ψ_L with density λ_2, as illustrated in Fig. 7.2. Further, we model the locations of the receiving vehicular nodes by a PLCP Φ_r on the same PLP Φ_l such that they form an independent and homogeneous 1D PPP with density λ_r on each line. We assume that the typical receiver is located at the origin. Thus, the spatial distribution of the receiving vehicular nodes is $\Phi_{r_0} \equiv \Phi_r \cup \Xi_{L_0} \cup \{o\}$, where Ξ_{L_0} represents the 1D PPP of receiving vehicular nodes located on the line L_0 passing through the origin, which will henceforth be referred to as the typical line. As a result, the locations of the transmitting tier 2 nodes under the Palm distribution of receiving vehicular nodes is given by $\Phi_2 \equiv \Phi_v \cup \Psi_{L_0}$. We assume that the tier 1 and tier 2 nodes have transmit powers P_1 and P_2, respectively. We denote the channel fading gains for the links from the typical receiver to the tier 1 and tier 2 nodes by H_1 and H_2, respectively.

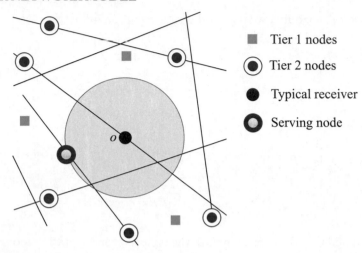

Figure 7.2: Illustration of the system model.

As the typical receiver connects to the node that yields the maximum average received power, the candidate serving nodes from tier 1 and tier 2 are the closest nodes to the typical receiver in their respective tiers. Note that under the Palm distribution of the receiving vehicular nodes, the typical line L_0 passing through the origin is deterministic (although we do not assume anything about the orientation of L_0), whereas the spatial distribution of other lines is random. Therefore, it is necessary to distinguish the two cases where the serving tier 2 node is on the typical line L_0 and the serving tier 2 node is located on any other line of the PLP. This will be discussed more elaborately in the next subsection. The locations of the candidate serving nodes from the set of tier 1 nodes, tier 2 nodes on the typical line, and tier 2 nodes located on the other lines are denoted by $x_1^* = \arg\min_{x \in \Phi_1} \|x\|$, $x_{20}^* = \arg\min_{x \in \Psi_{L_0}} \|x\|$, and $x_{21}^* = \arg\min_{x \in \Phi_2 \setminus \Psi_{L_0}} \|x\|$. We further assume a selection bias of B_1 and B_2 for the tier 1 and tier 2 nodes, respectively, to balance the load across the two tiers in the network. Thus, the location of the serving node is given by $x^* = \arg\min_{x \in \{x_j^*\}} P_j B_j \|x\|^{-\alpha}$, where $j \in \{1, 20, 21\}$, $P_{20} = P_{21} = P_2$, and $B_{20} = B_{21} = B_2$.

7.2.2   COVERAGE ANALYSIS

In this subsection, we will present a detailed derivation of the SINR-based coverage probability of the typical receiver. We will first lay out the key differences and technical challenges that emanate from the doubly stochastic nature of the PLCP model as compared to the canonical setup discussed in Section 7.1. First, recall that the spatial distribution of the interfering nodes in the cellular network model is a function of the distance to the serving node. Primarily, conditioning on the serving distance precludes the existence of interfering nodes within a certain distance from the typical receiver. For the setup discussed in the previous subsection, the transmitting tier 2 nodes are located on the lines of the underlying PLP and the serving node could be located on any of these lines. Consequently, the interference at the typical receiver depends on the line on which the serving node is located. Moreover, since the locations of nodes are coupled with the lines of the PLP, conditioning on the serving distance (the fact that there are no nodes closer than the serving distance on any of the lines) also affects the spatial distribution of the lines. This coupling needs to be carefully handled while computing the aggregate interference at the typical receiver.

Two different approaches have been proposed in the literature for the SINR-based coverage analysis of the cellular network model involving PLCP distribution of nodes [35, 40]. In [35], the coverage probability is computed by characterizing the interference measured at the typical receiver when the serving node is located on the nth closest line to the typical receiver. This intuitive approach to compute coverage probability is accomplished by leveraging the various distance distributions of the PLCP. The second approach to handle the coupling between the lines and the nodes is to derive the intermediate results by conditioning on the underlying PLP and obtain the final result by calculating the expectation w.r.t. the PLP in the last step, as proposed in [40]. We will present the coverage analysis based on the second approach in this section. While we will broadly follow the same analytical procedure for computing the coverage probability as presented in Fig. 7.1, some of the intermediate results corresponding to the tier 2 nodes will be derived by conditioning on the PLP Φ_l.

Step 1: Relevant distance distributions. We denote the distance from the typical receiver to the candidate serving nodes x_1^*, x_{20}^*, and x_{21}^* by R_1, R_{20}, and R_{21}, respectively. As the tier 1 nodes are modeled by a homogeneous 2D PPP, the CDF and PDF of R_1 are given by

$$F_{R_1}(r_1) = 1 - \exp(-\lambda_1 \pi r_1^2), \tag{7.17}$$

$$f_{R_1}(r_1) = 2\pi \lambda_1 r_1 \exp(-\lambda_1 \pi r_1^2). \tag{7.18}$$

Since the locations of the tier 2 nodes on the typical line follow a 1D PPP, the CDF and PDF of R_{20} are

$$F_{R_{20}}(r_{20}) = 1 - \exp(-\lambda_2 2 r_{20}), \tag{7.19}$$

$$f_{R_{20}}(r_{20}) = 2\lambda_2 \exp(-\lambda_2 2 r_{20}). \tag{7.20}$$

Using the contact distribution function of the PLCP presented in Lemma 4.8, the CDF and PDF of R_{21} are given by

$$F_{R_{21}}(r_{21}) = 1 - \exp\left[-2\pi\lambda_l \int_0^{r_{21}} \left(1 - \exp(-2\lambda_2 \sqrt{r_{21}^2 - \rho^2})\right) d\rho\right]. \tag{7.21}$$

$$f_{R_{21}}(r_{21}) = \left[2\pi\lambda_l \int_0^{r_{21}} \frac{2\lambda_2 r_{21}}{\sqrt{r_{21}^2 - \rho^2}} e^{-2\lambda_2 \sqrt{r_{21}^2 - \rho^2}} d\rho\right]$$
$$\times \exp\left[-2\pi\lambda_l \int_0^{r_{21}} \left(1 - \exp(-2\lambda_2 \sqrt{r_{21}^2 - \rho^2})\right) d\rho\right]. \tag{7.22}$$

Step 2: Association probabilities. We denote the events in which the typical receiver connects to tier 1 node, tier 2 node on the typical line, and tier 2 node on other lines by $\mathcal{E}_1$, $\mathcal{E}_{20}$, and $\mathcal{E}_{21}$, respectively. We will now compute the probability of occurrence of these events. We know that the event $\mathcal{E}_1$ would occur when the biased average received power from the candidate tier 1 node exceeds that of the candidate tier 2 nodes. Thus, the probability of occurrence of the event $\mathcal{E}_1$ can be computed as

$$
\mathbb{P}(\mathcal{E}_1) = \mathbb{P}\left(P_1 B_1 R_1^{-\alpha} > P_{20} B_{20} R_{20}^{-\alpha},\ P_1 B_1 R_1^{-\alpha} > P_{21} B_{21} R_{21}^{-\alpha}\right)
$$

$$
= \mathbb{P}\left(R_1 < \left(\frac{P_2 B_2}{P_1 B_1}\right)^{-\frac{1}{\alpha}} R_{20},\ R_1 < \left(\frac{P_2 B_2}{P_1 B_1}\right)^{-\frac{1}{\alpha}} R_{20}\right)
$$

$$
= \mathbb{E}_{R_1}\left[\mathbb{P}\left(R_{20} > \frac{r_1}{\zeta},\ R_{21} > \frac{r_1}{\zeta}\ \middle|\ R_1\right)\right],
$$

where $\zeta = \left(\frac{P_2 B_2}{P_1 B_1}\right)^{-\frac{1}{\alpha}}$. As R_{20} and R_{21} are independent, the probability of occurrence of $\mathcal{E}_1$ is obtained as

$$
\mathbb{P}(\mathcal{E}_1) = \int_0^\infty \left(1 - F_{R_{20}}\left(\frac{r_1}{\zeta}\right)\right)\left(1 - F_{R_{21}}\left(\frac{r_1}{\zeta}\right)\right) f_{R_1}(r_1)\,\mathrm{d}r_1
$$

$$
= \int_0^\infty \exp\left[-\lambda_1 \pi r_1^2 - 2\lambda_2 \frac{r_1}{\zeta} - 2\pi\lambda_l \int_0^{\frac{r_1}{\zeta}} 1 - \exp\left(-2\lambda_2 \sqrt{\frac{r_1^2}{\zeta^2} - \rho^2}\right)\mathrm{d}\rho\right] 2\pi\lambda_1 r_1\,\mathrm{d}r_1.
$$

$$(7.23)$$

Similarly, the probability of occurrence of the events $\mathcal{E}_{20}$ is given by

$$
\mathbb{P}(\mathcal{E}_{20}) = \int_0^\infty \exp\left[-\lambda_1 \pi \zeta^2 r_{20}^2 - 2\pi\lambda_l \int_0^{r_{20}} 1 - \exp\left(-2\lambda_2 \sqrt{r_{20}^2 - \rho^2}\right)\mathrm{d}\rho - 2\lambda_2 r_{20}\right] 2\lambda_2\,\mathrm{d}r_{20}.
$$

$$(7.24)$$

As the probabilities of events $\mathcal{E}_1$, $\mathcal{E}_{20}$, and $\mathcal{E}_{21}$ sum to 1, the probability of occurrence of $\mathcal{E}_{21}$ can be simply obtained as

$$
\mathbb{P}(\mathcal{E}_{21}) = 1 - \mathbb{P}(\mathcal{E}_0) - \mathbb{P}(\mathcal{E}_{20}).
$$

$$(7.25)$$

Step 3: Serving distance distributions. We will now determine the distribution of the serving distance R conditioned on the events $\mathcal{E}_1$, $\mathcal{E}_{20}$, and $\mathcal{E}_{21}$. The CDF of R conditioned on $\mathcal{E}_1$ can be computed as

$$
F_R(r \mid \mathcal{E}_1) = 1 - \mathbb{P}(R > r \mid \mathcal{E}_1) = 1 - \frac{\mathbb{P}(R > r, \mathcal{E}_1)}{\mathbb{P}(\mathcal{E}_1)}.
$$

From the condition for the occurrence of the event $\mathcal{E}_1$, it follows that

$$
F_R(r \mid \mathcal{E}_1) = 1 - \frac{\mathbb{P}(R_1 > r,\ R_1 < \min\{\zeta R_{20}, \zeta R_{21}\})}{\mathbb{P}(\mathcal{E}_1)}.
$$

Let $W = \min\{\zeta R_{20}, \zeta R_{21}\}$. Since R_{20} and R_{21} are independent, the CDF of W is given by

$$F_W(w) = 1 - \left(1 - F_{R_{20}}\left(\frac{w}{\zeta}\right)\right)\left(1 - F_{R_{21}}\left(\frac{w}{\zeta}\right)\right). \tag{7.26}$$

The CDF of R conditioned on $\mathcal{E}_1$ can now be written as

$$F_R(r \mid \mathcal{E}_1) = 1 - \frac{1}{\mathbb{P}(\mathcal{E}_1)}\mathbb{E}_W\left[\mathbb{P}\left(r < R_1 < w \mid W\right)\right]$$

$$= 1 - \frac{1}{\mathbb{P}(\mathcal{E}_1)}\int_r^\infty \left(F_{R_1}(w) - F_{R_1}(r)\right)f_W(w)\mathrm{d}w. \tag{7.27}$$

By taking the derivative of the above expression w.r.t. r, we obtain the PDF of R conditioned on $\mathcal{E}_1$ as

$$f_R(r \mid \mathcal{E}_1) = \frac{1}{\mathbb{P}(\mathcal{E}_1)}\int_r^\infty f_{R_1}(r)f_W(w)\mathrm{d}w$$

$$= \frac{1}{\mathbb{P}(\mathcal{E}_1)}f_{R_1}(r)\left(1 - F_W(r)\right)$$

$$= \frac{1}{\mathbb{P}(\mathcal{E}_1)}f_{R_1}(r)\left(1 - F_{R_{20}}\left(\frac{r}{\zeta}\right)\right)\left(1 - F_{R_{21}}\left(\frac{r}{\zeta}\right)\right). \tag{7.28}$$

Substituting the expressions for $F_{R_{20}}(\cdot)$ and $F_{R_{21}}(\cdot)$ in the above equation, we obtain the final expression for the PDF of R conditioned on $\mathcal{E}_1$ as

$$f_R(r \mid \mathcal{E}_1) = \frac{1}{\mathbb{P}(\mathcal{E}_1)}\exp\left[-\lambda_1\pi r^2 - 2\lambda_2\frac{r}{\zeta} - 2\pi\lambda_l\int_0^{\frac{r}{\zeta}}1 - \exp\left(-2\lambda_2\sqrt{\frac{r^2}{\zeta^2} - \rho^2}\right)\mathrm{d}\rho\right]2\pi\lambda_1 r. \tag{7.29}$$

Following the same procedure, we will now proceed to derive the conditional serving distance distribution for the case of $\mathcal{E}_{20}$. As mentioned earlier, we will compute this result by additionally conditioning on the underlying PLP Φ_l. Thus, the CDF of R conditioned on $\mathcal{E}_{20}$ and Φ_l is

$$F_R(r \mid \mathcal{E}_{20}, \Phi_l) = 1 - \mathbb{P}(R > r \mid \mathcal{E}_{20}, \Phi_l) = 1 - \frac{\mathbb{P}(R > r, \mathcal{E}_{20} \mid \Phi_l)}{\mathbb{P}(\mathcal{E}_{20} \mid \Phi_l)}.$$

From the condition for the occurrence of the event $\mathcal{E}_{20}$, it follows that

$$F_R(r \mid \mathcal{E}_{20}, \Phi_l) = 1 - \frac{\mathbb{P}(R_{20} > r, R_{20} < \min\{\zeta^{-1}R_1, R_{21}\} \mid \Phi_l)}{\mathbb{P}(\mathcal{E}_{20} \mid \Phi_l)}.$$

Let $V = \min\{\zeta^{-1}R_1, R_{21}\}$. Since R_1 and R_{21} are independent, the CDF of V conditioned on Φ_l is given by

$$F_V(v \mid \Phi_l) = 1 - \left(1 - F_{R_1}(\zeta v \mid \Phi_l)\right)\left(1 - F_{R_{21}}(v \mid \Phi_l)\right)$$
$$= 1 - \exp(-\lambda_1 \pi \zeta^2 v^2) \prod_{\substack{\rho \in \Phi_l \\ \rho < v}} \exp\left(-2\lambda_2\sqrt{v^2 - \rho^2}\right). \qquad (7.30)$$

The CDF of R conditioned on $\mathcal{E}_{20}$ can now be written as

$$F_R(r \mid \mathcal{E}_{20}, \Phi_l) = 1 - \frac{1}{\mathbb{P}(\mathcal{E}_{20} \mid \Phi_l)}\mathbb{E}_V\left[\mathbb{P}\left(r < R_{20} < v \mid V, \Phi_l\right)\right]$$
$$= 1 - \frac{1}{\mathbb{P}(\mathcal{E}_{20} \mid \Phi_l)}\int_r^\infty \left(F_{R_{20}}(v) - F_{R_{20}}(r)\right) f_V(v \mid \Phi_l)\mathrm{d}v. \qquad (7.31)$$

By taking the derivative of the above expression w.r.t. r, we obtain the PDF of the serving distance R conditioned on $\mathcal{E}_{20}$ and Φ_l as

$$f_R(r \mid \mathcal{E}_{20}, \Phi_l) = \frac{1}{\mathbb{P}(\mathcal{E}_{20} \mid \Phi_l)}\int_r^\infty f_{R_{20}}(r) f_V(v \mid \Phi_l)\mathrm{d}v$$
$$= \frac{1}{\mathbb{P}(\mathcal{E}_{20} \mid \Phi_l)}f_{R_{20}}(r)\left(1 - F_V(r \mid \Phi_l)\right). \qquad (7.32)$$

Substituting the expressions for $f_{R_{20}}(\cdot)$ and $F_V(\cdot \mid \Phi_l)$ in the above equation, we obtain the final expression for the distribution of the serving distance conditioned on $\mathcal{E}_{20}$ as

$$f_R(r \mid \mathcal{E}_{20}, \Phi_l) = \frac{1}{\mathbb{P}(\mathcal{E}_{20} \mid \Phi_l)}2\lambda_2 \exp\left[-2\lambda_2 r - \lambda_1 \pi \zeta^2 r^2\right]\prod_{\substack{\rho \in \Phi_l \\ \rho < r}} \exp\left(-2\lambda_2\sqrt{r^2 - \rho^2}\right).$$
$$(7.33)$$

We will now derive the conditional PDF of the serving distance for the case of $\mathcal{E}_{21}$. Let us assume that the serving node is located on the line $L_s \in \Phi_l$ which is represented by the point (ρ_s, θ_s) in the representation space. As will be shown later, while computing the interference at the typical receiver for this case, we will consider the interference caused by the nodes located on the line containing the serving node separately. Therefore, we will now derive the PDF of the serving distance R conditioned on $\mathcal{E}_{21}$, Φ_l and also the line L_s. First, the conditional CDF of R can be computed as

$$F_R(r \mid \mathcal{E}_{21}, \Phi_l, L_s) = 1 - \mathbb{P}\left(R > r \mid \mathcal{E}_{21}, \Phi_l, L_s\right).$$

Usually, conditioning on the event $\mathcal{E}_{21}$ implies that the serving distance R is the same as the distance R_{21} from the typical receiver to the closest node of the PLCP excluding the nodes on the typical line. However, note that we need to compute the serving distance distribution by additionally conditioning on the PLP Φ_l and the line containing the serving node L_s. Conditioning on the line L_s implies that the serving distance R is nothing but the distance from the typical receiver to its closest node on the line L_s. Let us denote this distance by R_s whose conditional CDF is given by

$$F_{R_s}(r_s \mid \Phi_l, L_s) = 1 - \mathbb{P}(R_s > r_s \mid \Phi_l, L_s) = 1 - \exp(-2\lambda_2 \sqrt{r_s^2 - \rho_s^2}). \qquad (7.34)$$

Thus, the CDF of R conditioned on $\mathcal{E}_{21}$, Φ_l, and L_s is

$$
\begin{aligned}
F_R(r \mid \mathcal{E}_{21}, \Phi_l, L_s) &= 1 - \mathbb{P}(R_s > r \mid \mathcal{E}_{21}, \Phi_l, L_s) \\
&= 1 - \frac{\mathbb{P}(R_s > r, \mathcal{E}_{21} \mid \Phi_l, L_s)}{\mathbb{P}(\mathcal{E}_{21} \mid \Phi_l, L_s)} \\
&= 1 - \frac{\mathbb{P}(R_s > r, R_s \leq \min\{\zeta^{-1}R_1, R_{20}, R_{21}\} \mid \Phi_l, L_s)}{\mathbb{P}(\mathcal{E}_{21} \mid \Phi_l, L_s)}.
\end{aligned}
\qquad (7.35)
$$

Let $U = \min\{\zeta^{-1}R_1, R_{20}, R_{21}\}$. Since R_1, R_{20}, and R_{21} are independent, the conditional CDF of U is given by

$$
\begin{aligned}
F_U(u \mid \Phi_l, L_s) &= 1 - \left(1 - F_{R_1}(\zeta u \mid \Phi_l, L_s)\right)\left(1 - F_{R_{20}}(u \mid \Phi_l, L_s)\right)\left(1 - F_{R_{21}}(u \mid \Phi_l, L_s)\right) \\
&= 1 - \exp\left[-\lambda_1 \pi \zeta^2 u^2 - 2\lambda_2 u\right] \prod_{\substack{\rho \in \Phi_l \\ \rho < u}} \exp\left(-2\lambda_2 \sqrt{u^2 - \rho^2}\right).
\end{aligned}
\qquad (7.36)
$$

Thus, we have

$$
\begin{aligned}
F_R(r \mid \mathcal{E}_{21}, \Phi_l, L_s) &= 1 - \frac{\mathbb{P}(r < R_s \leq u \mid U, \Phi_l, L_s)}{\mathbb{P}(\mathcal{E}_{21} \mid \Phi_l, L_s)} \\
&= 1 - \frac{1}{\mathbb{P}(\mathcal{E}_{21} \mid \Phi_l, L_s)} \int_r^\infty (F_{R_s}(u \mid \Phi_l, L_s) \\
&\qquad - F_{R_s}(r \mid \Phi_l, L_s)) f_U(u \mid \Phi_l, L_s) \, du.
\end{aligned}
\qquad (7.37)
$$

By taking the derivative of the above equation, we obtain the PDF of R conditioned on $\mathcal{E}_{21}$, Φ_l, and L_s as

$$
\begin{aligned}
f_R(r \mid \mathcal{E}_{21}, \Phi_l, L_s) &= \frac{1}{\mathbb{P}(\mathcal{E}_{21} \mid \Phi_l, L_s)} \int_r^\infty f_{R_s}(r \mid \Phi_l, L_s) f_U(u \mid \Phi_l, L_s) \, du \\
&= \frac{1}{\mathbb{P}(\mathcal{E}_{21} \mid \Phi_l, L_s)} f_{R_s}(r \mid \Phi_l, L_s)(1 - F_U(r \mid \Phi_l, L_s)) \\
&= \frac{1}{\mathbb{P}(\mathcal{E}_{21} \mid \Phi_l, L_s)} \frac{2\lambda_2 r \exp(-2\lambda_2 \sqrt{r^2 - \rho_s^2})}{\sqrt{r^2 - \rho_s^2}} \exp\left[-\lambda_1 \pi \zeta^2 r^2 - 2\lambda_2 r\right] \\
&\qquad \times \prod_{\substack{\rho \in \Phi_l \\ \rho < r}} \exp\left(-2\lambda_2 \sqrt{r^2 - \rho^2}\right).
\end{aligned}
\qquad (7.38)
$$

Table 7.1: Summary of the interference components under different conditions

Condition	Interference Components
$R, \mathcal{E}_1$	$I_1 : \mathrm{x} \in \Phi_1 \setminus B(o, r)$
	$I_{20}: \mathrm{x} \in \Psi_{L_0} \setminus B(o, \zeta^{-1}r)$
	$I_{21}: \mathrm{x} \in \{\Phi_2 \setminus \Psi_{L_0}\} \setminus B(o, \zeta^{-1}r); \rho < \zeta^{-1}r$
	$I_{22}: \mathrm{x} \in \Phi_2 \setminus \Psi_{L_0}; \rho > \zeta^{-1}r$
$R, \mathcal{E}_{20}, \Phi_l$	$I_1 : \mathrm{x} \in \Phi_1 \setminus B(o, \zeta r)$
	$I_{20}: \mathrm{x} \in \Psi_{L_0} \setminus B(o, r)$
	$I_{21}: \mathrm{x} \in \{\Phi_2 \setminus \Psi_{L0}\} \setminus B(o, r); \rho < r$
	$I_{22}: \mathrm{x} \in \Phi_2 \setminus \Psi_{L_0}; \rho > r$
$R, \mathcal{E}_{21}, \Phi_l, L_s$	$I_1 : \mathrm{x} \in \Phi_1 \setminus B(o, \zeta r)$
	$I_{20}: \mathrm{x} \in \Psi_{L_0} \setminus B(o, r)$
	$I_{21}: \mathrm{x} \in \{\Phi_2 \setminus \Psi_{L_0}\} \setminus B(o, r); \rho < r$
	$I_{22}: \mathrm{x} \in \Phi_2 \setminus \Psi_{L_0}; \rho > r$
	$I_{23}: \mathrm{x} \in \Psi_{L_s} \setminus B(o, r)$

Thus far, we have derived the serving distance distributions conditioned on the events $\mathcal{E}_1$, $\mathcal{E}_{20}$, and $\mathcal{E}_{21}$. The next key step is the characterization of aggregate interference at the typical receiver for these cases.

Step 4: Laplace transform of interference. The computation of conditional Laplace transform of interference is not quite straightforward in this model and is the main technical challenge involved in the computation of coverage probability. We need to compute the Laplace transform of interference power distribution corresponding to the cases $\mathcal{E}_1$, $\mathcal{E}_{20}$, and $\mathcal{E}_{21}$ and the restrictions on the locations of the interfering nodes is different for each case. Therefore, we summarize the locations of interfering nodes for each component under different conditions in Table 7.1. We will first focus on the Laplace transform of the distribution of the interference power conditioned on the serving distance R and the event $\mathcal{E}_1$. The interference from the tier 1 and tier 2 nodes, denoted by I_1 and I_2, are independent of each other and hence, will be handled separately. Conditioning on R and $\mathcal{E}_1$ implies that there are no interfering tier 1 nodes in the disc $B(o, R)$. Thus, the conditional Laplace transform of I_1 is given by

$$\mathcal{L}_{I_1}(s \mid \mathcal{E}_1) = \exp\left(-2\pi\lambda_1 \int_r^\infty \frac{sP_1 x^{1-\alpha}}{1 + sP_1 x^{-\alpha}} \mathrm{d}x\right). \qquad (7.39)$$

We will now focus on the interference caused by the tier 2 nodes. Conditioning on R and $\mathcal{E}_1$ implies that there are no tier 2 nodes inside the disc $B(o, \zeta^{-1}R)$. As the tier 2 nodes are located on the lines of the PLP Φ_{l_0}, the lines that intersect the disc $B(o, \zeta^{-1}R)$ cannot contain any nodes in the chord segment inside the disc. We know that the typical line L_0 passing through the origin always intersects the disc $B(o, \zeta^{-1}R)$. In addition to L_0, there are also a random number of lines that intersect $B(o, \zeta^{-1}R)$. The interfering nodes located on these lines are present only outside the disc $B(o, \zeta^{-1}R)$. The interfering tier 2 nodes are also located on the lines that do not intersect this disc. Thus, the interference from the tier 2 nodes can be decomposed into three components: (i) the interference from the nodes located on the typical line I_{20}; (ii) the interference from the nodes located on the lines that intersect the disc $B(o, \zeta^{-1}R)$ excluding the typical line, I_{21}; and (iii) the interference from the nodes located on the lines that do not intersect the disc $B(o, \zeta^{-1}R)$, I_{22}. Also, these three components of interference are independent and will be characterized separately. First, the Laplace transform of I_{20} can be easily computed from the Laplace functional of 1D PPP as

$$\mathcal{L}_{I_{20}}(s \mid \mathcal{E}_1) = \exp\left(-2\lambda_2 \int_{\zeta^{-1}r}^{\infty} \frac{sP_2 x^{-\alpha}}{1 + sP_2 x^{-\alpha}} \mathrm{d}x\right). \tag{7.40}$$

The Laplace transform of I_{21} conditioned on R and $\mathcal{E}_1$ can be computed as

$$\mathcal{L}_{I_{21}}(s \mid \mathcal{E}_1) = \mathbb{E}\left[\exp\left(-s \sum_{\substack{x \in \Phi_2 \setminus \Psi_{L_0} \\ \rho < \zeta^{-1}r}} P_2 H_2 \|x\|^{-\alpha}\right)\right]$$

$$= \mathbb{E}_{\Phi_2}\left[\prod_{\substack{x \in \Phi_2 \setminus \Psi_{L_0} \\ \rho < \zeta^{-1}r}} \frac{1}{1 + sP_2 \|x\|^{-\alpha}}\right]. \tag{7.41}$$

For a given line at a distance ρ that intersects the disc $B(o, \zeta^{-1}r)$, there cannot be any point within a distance $\sqrt{\zeta^{-2}r^2 - \rho^2}$ from the foot of the perpendicular on the line. Using the Laplace functional of the PLCP presented in Section 4.3 and incorporating the above restriction on the location of the nodes, we obtain

$$\mathcal{L}_{I_{21}}(s \mid \mathcal{E}_1) = \exp\left[-2\pi\lambda_l \int_0^{\zeta^{-1}r} 1 - \exp\left[-2\lambda_2 \int_{\sqrt{\zeta^{-2}r^2 - \rho^2}}^{\infty} \frac{P_2 s}{P_2 s + (\rho^2 + u^2)^{\alpha/2}} \mathrm{d}u\right] \mathrm{d}\rho\right]. \tag{7.42}$$

We will now compute the Laplace transform of the distribution of interference power due to the nodes located on the lines whose distance from the origin is greater than $\zeta^{-1}r$. Note that the conditioning on $\mathcal{E}_1$ does not impose any restrictions on the locations of the nodes on these lines. Thus, we have

$$\mathcal{L}_{I_{22}}(s \mid \mathcal{E}_1) = \exp\left[-2\pi\lambda_l \int_{\zeta^{-1}r}^{\infty} 1 - \exp\left[-2\lambda_2 \int_0^{\infty} \frac{P_2 s}{P_2 s + (\rho^2 + u^2)^{\alpha/2}} du \right] d\rho \right]. \tag{7.43}$$

Thus, the Laplace transform of interference from all the tier 2 nodes conditioned on the serving distance R and the event $\mathcal{E}_1$ can be computed as

$$\mathcal{L}_{I_2}(s \mid \mathcal{E}_1) = \mathcal{L}_{I_{20}}(s \mid \mathcal{E}_1)\mathcal{L}_{I_{21}}(s \mid \mathcal{E}_1)\mathcal{L}_{I_{22}}(s \mid \mathcal{E}_1). \tag{7.44}$$

The Laplace transform of the distribution of aggregate interference at the typical receiver conditioned on R and $\mathcal{E}_1$ is given by

$$\mathcal{L}_I(s \mid \mathcal{E}_1) = \mathcal{L}_{I_1}(s \mid \mathcal{E}_1)\mathcal{L}_{I_2}(s \mid \mathcal{E}_1). \tag{7.45}$$

Substituting (7.39), (7.40), (7.42), (7.43), and (7.44) in the above equation, we obtain

$$\mathcal{L}_I(s \mid \mathcal{E}_1) = \exp\left[-2\pi\lambda_1 \int_r^{\infty} \frac{sP_1 x^{1-\alpha}}{1 + sP_1 x^{-\alpha}} dx - 2\lambda_2 \int_{\zeta^{-1}r}^{\infty} \frac{sP_2 x^{-\alpha}}{1 + sP_2 x^{-\alpha}} dx \right.$$
$$- 2\pi\lambda_l \int_0^{\zeta^{-1}r} 1 - \exp\left[-2\lambda_2 \int_{\sqrt{\zeta^{-2}r^2 - \rho^2}}^{\infty} \frac{P_2 s}{P_2 s + (\rho^2 + u^2)^{\alpha/2}} du \right] d\rho$$
$$\left. - 2\pi\lambda_l \int_{\zeta^{-1}r}^{\infty} 1 - \exp\left[-2\lambda_2 \int_0^{\infty} \frac{P_2 s}{P_2 s + (\rho^2 + u^2)^{\alpha/2}} du \right] d\rho \right]. \tag{7.46}$$

Following the same procedure, we will now compute the Laplace transform of the interference power distribution conditioned on R and $\mathcal{E}_{20}$. We will calculate the Laplace transform of independent components of interference I_1, I_{20}, I_{21}, and I_{22} separately. Conditioning on R and $\mathcal{E}_{20}$ implies that there are no interfering tier 1 nodes in the disc $B(o, \zeta R)$. Thus, the conditional Laplace transform of I_1 is given by

$$\mathcal{L}_{I_1}(s \mid \mathcal{E}_{20}) = \exp\left(-2\pi\lambda_1 \int_{\zeta r}^{\infty} \frac{sP_1 x^{1-\alpha}}{1 + sP_1 x^{-\alpha}} dx \right). \tag{7.47}$$

We will now compute the Laplace transform of interference from tier 2 nodes by additionally conditioning on the PLP Φ_l. First, the Laplace transform of I_{20} can be easily computed from the Laplace functional of 1D PPP as

$$\mathcal{L}_{I_{20}}(s \mid \mathcal{E}_{20}, \Phi_l) = \exp\left(-2\lambda_2 \int_r^{\infty} \frac{sP_2 x^{-\alpha}}{1 + sP_2 x^{-\alpha}} dx \right). \tag{7.48}$$

The Laplace transform of I_{21} conditioned on R and $\mathcal{E}_{20}$ can be computed as

$$\mathcal{L}_{I_{21}}(s \mid \mathcal{E}_{20}, \Phi_l) = \mathbb{E}\left[\exp\left(-s \sum_{\substack{x\in\Phi_2\backslash\Psi_{L_0} \\ \rho<r}} P_2 H_2 \|x\|^{-\alpha}\right)\right]$$

$$= \mathbb{E}_{\Phi_{20}}\left[\prod_{\substack{x\in\Phi_2\backslash\Psi_{L_0} \\ \rho<r}} \frac{1}{1 + sP_2\|x\|^{-\alpha}}\right]$$

$$= \prod_{\substack{\rho\in\Phi_l \\ \rho<r}} \exp\left[-2\lambda_2 \int_{\sqrt{r^2-\rho^2}}^{\infty} \frac{P_2 s}{P_2 s + (\rho^2 + u^2)^{\alpha/2}} du\right]. \tag{7.49}$$

We will now compute the Laplace transform of the distribution of interference power due to the nodes located on the lines whose distance from the origin is greater than r. As the conditioning on the event $\mathcal{E}_{20}$ does not impose any restrictions on the locations of the nodes on these lines, we have

$$\mathcal{L}_{I_{22}}(s \mid \mathcal{E}_{20}, \Phi_l) = \prod_{\substack{\rho\in\Phi_l \\ \rho>r}} \exp\left[-2\lambda_2 \int_0^{\infty} \frac{P_2 s}{P_2 s + (\rho^2 + u^2)^{\alpha/2}} du\right]. \tag{7.50}$$

Thus, the Laplace transform of the distribution of aggregate interference at the typical receiver conditioned on R, $\mathcal{E}_{20}$, and Φ_l is given by

$$\mathcal{L}_I(s \mid \mathcal{E}_{20}, \Phi_l) = \mathcal{L}_{I_1}(s \mid \mathcal{E}_{20}, \Phi_l)\mathcal{L}_{I_{20}}(s \mid \mathcal{E}_{20}, \Phi_l)$$
$$\times \mathcal{L}_{I_{21}}(s \mid \mathcal{E}_{20}, \Phi_l)\mathcal{L}_{I_{22}}(s \mid \mathcal{E}_{20}, \Phi_l). \tag{7.51}$$

Substituting (7.47), (7.48), (7.49), and (7.50) in the above equation, we obtain the Laplace transform of the distribution of the aggregate interference conditioned on R, $\mathcal{E}_{20}$, and Φ_l as

$$\mathcal{L}_I(s \mid \mathcal{E}_{20}, \Phi_l) = \exp\left(-2\pi\lambda_1 \int_{\xi r}^{\infty} \frac{sP_1 x^{1-\alpha}}{1 + sP_1 x^{-\alpha}} dx - 2\lambda_2 \int_r^{\infty} \frac{sP_2 x^{-\alpha}}{1 + sP_2 x^{-\alpha}} dx\right)$$

$$\times \prod_{\substack{\rho\in\Phi_l \\ \rho<r}} \exp\left[-2\lambda_2 \int_{\sqrt{r^2-\rho^2}}^{\infty} \frac{P_2 s}{P_2 s + (\rho^2 + u^2)^{\alpha/2}} du\right]$$

$$\times \prod_{\substack{\rho\in\Phi_l \\ \rho>r}} \exp\left[-2\lambda_2 \int_0^{\infty} \frac{P_2 s}{P_2 s + (\rho^2 + u^2)^{\alpha/2}} du\right]. \tag{7.52}$$

We will now characterize the interference at the typical receiver for the case of $\mathcal{E}_{21}$. Recall that the serving node in this case is assumed to be located on the line L_s which corresponds to (ρ_s, θ_s) in the representation space. In this case, we compute the Laplace transform of interference by additionally conditioning on the line L_s. Therefore, in addition to the independent interference components I_1, I_{20}, I_{21}, and I_{22}, we must also compute the interference caused by the nodes located on the line L_s, denoted by I_{23}, separately. Please note that the expressions for the Laplace transform of interference for other components (excluding I_{23}) are the same for the case of $\mathcal{E}_{20}$ and $\mathcal{E}_{21}$. So, we will just focus on the Laplace transform of I_{23} conditioned on R, $\mathcal{E}_{21}$ and L_s which can be computed as

$$
\begin{aligned}
\mathcal{L}_{I_{23}}(s \mid \mathcal{E}_{21}, \Phi_l, L_s) &= \mathbb{E}\left[\exp\left(-s \sum_{x \in \Psi_{L_s}} P_2 H_2 \|x\|^{-\alpha} \right) \right] \\
&= \mathbb{E}\left[\prod_{x \in \Psi_{L_s}} \exp\left(-s P_2 H_2 (\rho_s^2 + u^2)^{-\frac{\alpha}{2}} \right) \right] \\
&= \exp\left[-2\lambda_2 \int_{\sqrt{r^2-\rho_s^2}}^{\infty} \frac{P_2 s}{P_2 s + (\rho_s^2 + u^2)^{\frac{\alpha}{2}}} du \right].
\end{aligned} \tag{7.53}
$$

Using the above expression, the Laplace transform of the distribution of the aggregate interference at the typical receiver is given by

$$
\begin{aligned}
\mathcal{L}_I(s \mid \mathcal{E}_{21}, \Phi_l, L_s) &= \exp\left(-2\pi\lambda_1 \int_{\xi r}^{\infty} \frac{s P_1 x^{1-\alpha}}{1 + s P_1 x^{-\alpha}} dx - 2\lambda_2 \int_r^{\infty} \frac{s P_2 x^{-\alpha}}{1 + s P_2 x^{-\alpha}} dx \right) \\
&\quad \times \exp\left(-2\lambda_2 \int_{\sqrt{r^2-\rho_s^2}}^{\infty} \frac{P_2 s}{P_2 s + (\rho_s^2 + u^2)^{\frac{\alpha}{2}}} du \right) \\
&\quad \times \prod_{\substack{\rho \in \Phi_l \\ \rho < r}} \exp\left[-2\lambda_2 \int_{\sqrt{r^2-\rho^2}}^{\infty} \frac{P_2 s}{P_2 s + (\rho^2 + u^2)^{\alpha/2}} du \right] \\
&\quad \times \prod_{\substack{\rho \in \Phi_l \\ \rho > r}} \exp\left[-2\lambda_2 \int_0^{\infty} \frac{P_2 s}{P_2 s + (\rho^2 + u^2)^{\alpha/2}} du \right].
\end{aligned} \tag{7.54}
$$

Step 5: Coverage probability. We will now compute the coverage probability using law of total probability as follows:

$$
P_c = \mathbb{P}(\text{SINR} > \beta, \mathcal{E}_1) + \mathbb{P}(\text{SINR} > \beta, \mathcal{E}_{20}) + \mathbb{P}(\text{SINR} > \beta, \mathcal{E}_{21}). \tag{7.55}
$$

We will now present the exact expression for each of the terms in the above equation. First, we determine the coverage probability when the typical receiver connects to a tier 1 node as

$$
\mathbb{P}(\text{SINR} > \beta, \mathcal{E}_1) = \mathbb{P}(\mathcal{E}_1)\mathbb{P}(\text{SINR} > \beta \mid \mathcal{E}_1). \tag{7.56}
$$

Using (5.5), the coverage probability conditioned on the event $\mathcal{E}_1$ can be written as

$$\mathbb{P}(\text{SINR} > \beta \mid \mathcal{E}_1) = \int_0^\infty e^{-\frac{\beta r^\alpha}{P_1}\sigma^2} \mathcal{L}_I\left(\frac{\beta r^\alpha}{P_1} \mid \mathcal{E}_1\right) f_R(r \mid \mathcal{E}_1)\, dr. \tag{7.57}$$

Substituting (7.29), (7.46), and (7.57) in (7.56), we obtain

$$\mathbb{P}(\text{SINR} > \beta, \mathcal{E}_1)$$
$$= \int_0^\infty \exp\left[-\frac{\beta r^\alpha}{P_1}\sigma^2 - 2\pi\lambda_1 \int_r^\infty \frac{\beta r^\alpha x^{1-\alpha}}{1 + \beta r^\alpha x^{-\alpha}}\, dx - 2\lambda_2 \int_{\zeta^{-1}r}^\infty \frac{\beta r^\alpha P_2 x^{-\alpha}}{P_1 + \beta r^\alpha P_2 x^{-\alpha}}\, dx\right.$$
$$- 2\pi\lambda_l \int_0^{\zeta^{-1}r} 1 - \exp\left[-2\lambda_2 \int_{\sqrt{\zeta^{-2}r^2-\rho^2}}^\infty \frac{P_2\beta r^\alpha}{P_2\beta r^\alpha + P_1(\rho^2 + u^2)^{\alpha/2}}\, du\right] d\rho$$
$$- 2\pi\lambda_l \int_{\zeta^{-1}r}^\infty 1 - \exp\left[-2\lambda_2 \int_0^\infty \frac{P_2\beta r^\alpha}{P_2\beta r^\alpha + P_1(\rho^2 + u^2)^{\alpha/2}}\, du\right] d\rho$$
$$\left. - \lambda_1\pi r^2 - 2\lambda_2\zeta^{-1}r - 2\pi\lambda_l \int_0^{\zeta^{-1}r} 1 - \exp\left(-2\lambda_2\sqrt{\zeta^{-2}r^2 - \rho^2}\right) d\rho\right] 2\pi\lambda_1 r\, dr. \tag{7.58}$$

We will now compute the coverage probability when the typical receiver connects to the tier 2 node on the typical line as

$$\mathbb{P}(\text{SINR} > \beta, \mathcal{E}_{20}) = \mathbb{P}(\mathcal{E}_{20})\mathbb{P}(\text{SINR} > \beta \mid \mathcal{E}_{20}) = \mathbb{E}_{\Phi_l}\left[\mathbb{P}(\mathcal{E}_{20} \mid \Phi_l)\mathbb{P}(\text{SINR} > \beta \mid \mathcal{E}_{20}, \Phi_l)\right]. \tag{7.59}$$

Similar to the previous case, we express the coverage probability conditioned on $\mathcal{E}_{20}$ and Φ_l in terms of the conditional Laplace transform of interference and conditional serving distance distribution. Thus, we have

$$\mathbb{P}(\text{SINR} > \beta, \mathcal{E}_{20})$$

$$= \mathbb{E}_{\Phi_l}\left[\mathbb{P}(\mathcal{E}_{20} \mid \Phi_l)\int_0^\infty e^{-\frac{\beta r^\alpha}{P_2}\sigma^2}\mathcal{L}_I\left(\frac{\beta r^\alpha}{P_2} \mid \mathcal{E}_{20}, \Phi_l\right)f_R\left(r \mid \mathcal{E}_{20}, \Phi_l\right)dr\right]$$

$$= \mathbb{E}_{\Phi_l}\left[\int_0^\infty \exp\left(-\frac{\beta r^\alpha}{P_2}\sigma^2 - 2\pi\lambda_1\int_{\zeta r}^\infty \frac{\beta r^\alpha P_1 x^{1-\alpha}}{P_2 + \beta r^\alpha P_1 x^{-\alpha}}dx - 2\lambda_2\int_r^\infty \frac{\beta r^\alpha x^{-\alpha}}{1 + \beta r^\alpha x^{-\alpha}}dx\right)\right.$$

$$\times \prod_{\substack{\rho\in\Phi_l \\ \rho<r}}\exp\left[-2\lambda_2\int_{\sqrt{r^2-\rho^2}}^\infty \frac{\beta r^\alpha}{\beta r^\alpha + (\rho^2 + u^2)^{\alpha/2}}du\right]$$

$$\times \prod_{\substack{\rho\in\Phi_l \\ \rho>r}}\exp\left[-2\lambda_2\int_0^\infty \frac{\beta r^\alpha}{\beta r^\alpha + (\rho^2 + u^2)^{\alpha/2}}du\right]$$

$$\left.\times 2\lambda_2\exp\left[-2\lambda_2 r - \lambda_1\pi\zeta^2 r^2\right]\prod_{\substack{\rho\in\Phi_l \\ \rho<r}}\exp\left(-2\lambda_2\sqrt{r^2-\rho^2}\right)dr\right]. \tag{7.60}$$

By applying Fubini's theorem, we can change the order of integration in the above equation which allows us to evaluate the expectation of the integrand w.r.t. Φ_l. The expectation of the product terms can be simplified using PGFL of Φ_l and therefore we obtain the expression for coverage probability as

$$\mathbb{P}(\text{SINR} > \beta, \mathcal{E}_{20})$$

$$= \int_0^\infty \exp\left[-\frac{\beta r^\alpha}{P_2}\sigma^2 - 2\pi\lambda_1\int_{\zeta r}^\infty \frac{\beta r^\alpha P_1 x^{1-\alpha}}{P_2 + \beta r^\alpha P_1 x^{-\alpha}}dx - 2\lambda_2\int_r^\infty \frac{\beta r^\alpha x^{-\alpha}}{1 + \beta r^\alpha x^{-\alpha}}dx\right.$$

$$- 2\pi\lambda_l\int_0^r 1 - \exp\left(-2\lambda_2\sqrt{r^2-\rho^2} - 2\lambda_2\int_{\sqrt{r^2-\rho^2}}^\infty \frac{\beta r^\alpha}{\beta r^\alpha + (\rho^2 + u^2)^{\alpha/2}}du\right)d\rho$$

$$- 2\pi\lambda_l\int_r^\infty 1 - \exp\left(-2\lambda_2\int_0^\infty \frac{\beta r^\alpha}{\beta r^\alpha + (\rho^2 + u^2)^{\alpha/2}}du\right)d\rho\right]$$

$$2\lambda_2\exp\left[-2\lambda_2 r - \lambda_1\pi\zeta^2 r^2\right]dr. \tag{7.61}$$

Following the same procedure, we obtain the coverage probability of the typical receiver when it connects to tier 2 nodes on other lines as

$$
\mathbb{P}(\text{SINR} > \beta, \mathcal{E}_{21})
$$

$$
= \int_0^\infty \exp\left[-\frac{\beta r^\alpha}{P_2}\sigma^2 - 2\pi\lambda_1 \int_{\zeta r}^\infty \frac{\beta r^\alpha P_1 x^{1-\alpha}}{P_2 + \beta r^\alpha P_1 x^{-\alpha}}dx - 2\lambda_2 \int_r^\infty \frac{\beta r^\alpha x^{-\alpha}}{1 + \beta r^\alpha x^{-\alpha}}dx \right.
$$

$$
- 2\pi\lambda_l \int_0^r 1 - \exp\left(-2\lambda_2 \sqrt{r^2 - \rho^2} - 2\lambda_2 \int_{\sqrt{r^2 - \rho^2}}^\infty \frac{\beta r^\alpha}{\beta r^\alpha + (\rho^2 + u^2)^{\alpha/2}}du \right)d\rho
$$

$$
\left. - 2\pi\lambda_l \int_r^\infty 1 - \exp\left(-2\lambda_2 \int_0^\infty \frac{\beta r^\alpha}{\beta r^\alpha + (\rho^2 + u^2)^{\alpha/2}}du \right)d\rho \right]
$$

$$
\times \left(2\pi\lambda_l \int_0^r \exp\left(-2\lambda_2 \sqrt{r^2 - \rho_s^2} - 2\lambda_2 \int_{\sqrt{r^2 - \rho_s^2}}^\infty \frac{\beta r^\alpha}{\beta r^\alpha + (\rho_s^2 + u^2)^{\frac{\alpha}{2}}}du \right) \frac{d\rho_s}{\sqrt{r^2 - \rho_s^2}} \right)
$$

$$
\times 2\lambda_2 r \exp\left[-2\lambda_2 r - \lambda_1 \pi \zeta^2 r^2 \right]dr. \tag{7.62}
$$

Thus, we have derived the coverage probability of the typical receiver corresponding to the events $\mathcal{E}_1$, $\mathcal{E}_{20}$, and $\mathcal{E}_{21}$. We can obtain the expression for the overall coverage probability by substituting (7.58), (7.61), and (7.62) in (7.55).

7.3   PERFORMANCE TRENDS

In this section, we will analyze the trends in the SINR-based coverage probability of the typical receiver for the PLCP model. We plot the coverage probability as a function of density of tier 2 nodes for different values of line densities in Fig. 7.3. As the line density increases, the coverage probability of the typical receiver decreases due to an increase in the interference power. However, the coverage probability improves with the densification of the tier 2 nodes in the network. Upon increasing the density of tier 2 nodes, the distance to the closest tier 2 node on the typical line decreases and consequently, the desired signal power increases. At the same time, the nodes located on the same line come closer to each other. Since this does not directly translate to a reduced distance between the typical receiver and interfering nodes, the effect of increase in the interference power is not as dominant as the increment in the desired signal power. Hence, we observe an overall improvement in the coverage probability.

On the other hand, we do not observe a significant change in the coverage probability upon increasing the density of tier 1 nodes, as shown in Fig. 7.4. This suggests that the tier 2 nodes have a much higher contribution to the coverage probability than the tier 1 nodes. However, upon closer inspection, we still observe a decreasing trend in coverage probability as the density of the tier 1 nodes increases. An increase in the density of tier 1 nodes would increase the desired signal power and interference power proportionately when the typical receiver is connected to a

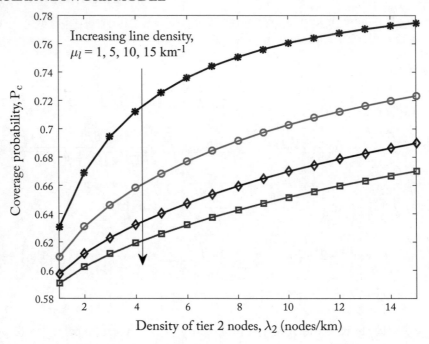

Figure 7.3: Coverage probability of the typical receiver as a function of tier 2 node density ($\lambda_1 = 0.5$ nodes/km^2, P_1 = 43 dBm, P_2 = 23 dBm, $\beta = 0$ dB, and $\alpha = 4$).

tier 1 node. However, when the typical receiver is connected to a tier 2 node, a higher density of tier 1 nodes would only increase the interference power at the typical receiver, thereby degrading the overall coverage probability, as depicted in Fig. 7.4.

7.4   SINR CHARACTERIZATION UNDER SHADOWING

Although we have provided a detailed derivation of the SINR-based coverage probability in Section 7.2.2, we did not consider the effect of shadowing in our system model for simplicity of exposition. However, in urban vehicular networks, the signals are often blocked by buildings and other structures and, hence, it is important to consider the impact of shadowing on the network performance. Therefore, in this section, we will discuss the computation of SINR with the inclusion of shadowing effects. We will first explain the technical challenges involved in the characterization of SINR in the presence of shadowing effects in a vehicular network modeled by a PLCP. We will then discuss an approximate spatial model to the PLCP that lends tractability to the analysis of this setup.

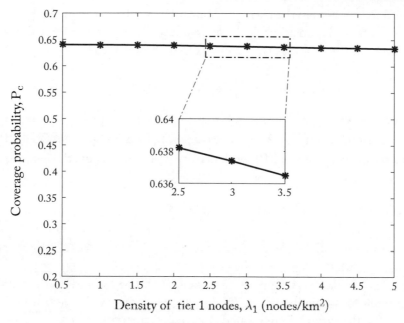

Figure 7.4: Coverage probability of the typical receiver as a function of tier 1 node density ($\mu_l = 10$ km^{-1}, $\lambda_2 = 5$ nodes/km, $P_1 = 43$ dBm, $P_2 = 23$ dBm, $\beta = 0$ dB, and $\alpha = 4$).

7.4.1   CHALLENGES DUE TO SHADOWING

In order to clearly understand the analytical challenges due to the inclusion of shadowing, let us consider the same vehicular network as in Section 7.2.1. However, additionally, we now model the effect of shadowing by log-normal random variables. Note that there is a higher likelihood of a line-of-sight (LoS) link between the typical receiver and tier 2 nodes on the typical line as compared to the tier 2 nodes on other lines. Therefore, it is important to distinguish the severity of shadowing between these two types of links. So, we denote the shadowing gains for the links between the typical receiver and the tier 1 nodes, tier 2 nodes on the typical line, and tier 2 nodes on the other lines by $\mathcal{X}_1$, $\mathcal{X}_{20}$, and $\mathcal{X}_{21}$, respectively. As mentioned earlier, the random variables $\mathcal{X}_k, k \in \{1, 20, 21\}$, follow log-normal distribution such that $10 \log_{10} \mathcal{X}_k \sim \mathcal{N}(\omega_k, \sigma_k^2)$ with mean ω_k and standard deviation σ_k in dB. We consider an association policy by which the typical receiver would connect to the node that yields the maximum average received power.

To compute the SINR at the typical receiver, the first step is to identify the candidate serving nodes from both tier 1 and tier 2 nodes. For a tier 1 node located at x $\in \Phi_1$, the average received power measured at the typical receiver is given by $P_1 \mathcal{X}_1 \|x\|^{-\alpha}$. Thus, the candidate serving node from tier 1 is the node located at

$$x_1^* = \arg \max_{x \in \Phi_1} P_1 \mathcal{X}_1 \|x\|^{-\alpha}. \tag{7.63}$$

The standard procedure followed in the literature to handle the effect of shadowing is to absorb it as an equivalent random displacement of the location of the node w.r.t. the typical receiver [84]. Therefore, the location of the candidate tier 1 node can be expressed as

$$x_1^* = \arg \max_{x \in \Phi_1} P_1 \|\mathcal{X}_1^{-\frac{1}{\alpha}} x\|^{-\alpha}. \tag{7.64}$$

From displacement theorem [6, 84], it follows that the new point process $\Omega' \equiv \{y\}$ obtained upon applying the transformation $y = \mathcal{X}^{-\frac{1}{\alpha}} x$ to each point of a homogeneous 2D PPP $\Omega \equiv \{x\}$ with density λ is also a homogeneous PPP with density $\mathbb{E}[\mathcal{X}^{-\frac{2}{\alpha}}]\lambda$. Thus, the candidate serving node from tier 1 is

$$x_1^* = \arg \max_{y \in \Phi_1'} P_1 \|y\|^{-\alpha}, \tag{7.65}$$

where Φ_1' is a homogeneous 2D PPP with density $\mu_1 = \mathbb{E}[\mathcal{X}_1^{-\frac{2}{\alpha}}]\lambda_1$. We will now examine this approach for the tier 2 nodes.

Recall that the candidate serving node from tier 2 could be located on the typical line or the other lines. Therefore, we will consider the candidate serving node from these two set of nodes separately. Thus, the candidate tier 2 node located on the typical line is

$$x_{20}^* = \arg \max_{x \in \Psi_{L_0}} P_2 \mathcal{X}_{20} \|x\|^{-\alpha}. \tag{7.66}$$

Applying the same technique as discussed earlier and expressing the effect of shadowing as a random displacement of the nodes w.r.t. the typical receiver, the location of candidate tier 2 node from the typical line is obtained as

$$x_{20}^* = \arg \max_{x \in \Psi_{L_0}} P_2 \|\mathcal{X}_{20}^{-\frac{1}{\alpha}} x\|^{-\alpha} = \arg \max_{y \in \Psi_{L_0}'} P_2 \|y\|^{-\alpha}, \tag{7.67}$$

where Ψ_{L_0}' is a homogeneous 1D PPP with density $\mu_2 = \mathbb{E}[\mathcal{X}_{20}^{-\frac{1}{\alpha}}]\lambda_2$.

Among the tier 2 nodes located on the other lines, the location of the candidate serving node is given by

$$x_{21}^* = \arg \max_{x \in \Phi_2 \backslash \Psi_{L_0}} P_2 \mathcal{X}_{21} \|x\|^{-\alpha}. \tag{7.68}$$

Upon expressing the shadowing effect as a random displacement of the nodes, we obtain

$$x_{21}^* = \arg \max_{x \in \Phi_2 \backslash \Psi_{L_0}} P_2 \|\mathcal{X}_{21}^{-\frac{1}{\alpha}} x\|^{-\alpha}. \tag{7.69}$$

It is important to note that the random displacement of the nodes $\mathcal{X}_{21}^{-\frac{1}{\alpha}} x$ is along the direction of the typical receiver, as illustrated in Fig. 7.5. As a result, the nodes that were located on the same line are not collinear anymore. Also, it is quite hard to determine the exact distribution of

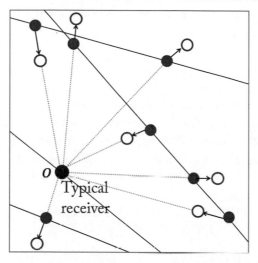

Figure 7.5: Illustration of the random displacement of points of a PLCP.

the new locations of the nodes. Hence, unlike the case of tier 1 nodes and tier 2 nodes on the typical line, this approach is not suitable for identifying the candidate serving node from tier 2 nodes on the other lines. We also encounter the same issue in characterizing the interference power from the tier 2 nodes located on the other lines. In order to overcome this problem, an approximation of the PLCP has been proposed in [37], which will be discussed next.

7.4.2   APPROXIMATION OF THE PLCP

From our discussion on the asymptotic characteristics of the PLCP in Section 4.4, we already know that the PLCP converges to a homogeneous 2D PPP as the line density tends to infinity and point density approaches zero while the average number of points per unit area remains unchanged. Moreover, we have shown that the random displacement of the points of the PLCP due to shadowing disrupts the collinear structure of all points with the exception of the points on the typical line. Inspired by these facts, the reduced Palm distribution of the PLCP can be approximated by the superposition of a homogeneous 1D PPP along the typical line and a homogeneous 2D PPP with the same intensity as the PLCP. Specifically, for the setup considered in the previous subsection, we approximate the PLCP Φ_2 by $\tilde{\Phi}_2 = \Psi'_{L_0} \cup \Phi_a$, where Ψ_{L_0} is a

1D PPP with density μ_2 and Φ_a is a 2D PPP with density $\mu_a = \mathbb{E}[\mathcal{X}_{21}^{-\frac{2}{\alpha}}]\pi\lambda_l\lambda_2$. This approximation significantly simplifies the spatial model and enables the analytical characterization of SINR at the typical receiver. Also, as will be shown later in the next subsection, this approximation works remarkably well even for moderate values of line and node densities. There are two key reasons behind the effectiveness of this approximate spatial model. First, from the perspective of mathematical modeling, we observe that the Ripley's K-function of the proposed model is the same as that of the PLCP. Secondly, the proposed approximate model allows us to accurately characterize the dominant interference field. We will discuss these two points in detail next.

Ripley's K-function. The reduced second moment function or Ripley's K-function is a metric that represents the characteristics of point processes at many distance scales and also identifies clustering, dispersion or complete spatial randomness exhibited by point patterns. We observe that the Ripley's K-function of the PLCP is identical to that of the proposed approximation. In order to establish this concretely, we will now present the derivation of the K-function for the motion-invariant PLCP Φ_2. For a motion invariant point process Ω, the Ripley's K-function is defined as the ratio of the mean number of points of Ω within a ball of radius r centered at the typical point (which is not counted) to the intensity of the point process. Recall that Φ_2 is the superposition of Φ_v and 1D PPP Ψ_{L_0} along the line L_0 passing through the origin. Thus, the mean number of points of the PLCP inside a ball of radius r centered at the origin is

$$
\begin{aligned}
\mathbb{E}\left[N_p(\Phi_2 \cap B(o,r))\right] &= \mathbb{E}\left[N_p\left((\Phi_v \cup \Psi_{L_0}) \cap B(o,r)\right)\right] \\
&= \mathbb{E}\left[N_p(\Psi_{L_0} \cap B(o,r))\right] + \mathbb{E}\left[N_p(\Phi_v \cap B(o,r))\right] \\
&= \lambda_p(2r) + \mu_l \lambda_p(\pi r^2),
\end{aligned}
\tag{7.70}
$$

where $N_p(A)$ denotes the number of points in set A. Thus, the K-function of the PLCP is computed as

$$
K(r) = \frac{1}{\lambda} \mathbb{E}\left[N_p(\Phi_2 \cap B(o,r))\right] = \frac{1}{\mu_l \lambda_p}\left(\lambda_p(2r) + \mu_l \lambda_p(\pi r^2)\right) = \frac{2r}{\mu_l} + \pi r^2.
\tag{7.71}
$$

The first term in the above equation corresponds to the points on the typical line L_0. The second term πr^2 in the K-function, which corresponds to the points on the other lines of the PLCP, is the same as the K-function of a homogeneous 2D PPP. This further shows that the proposed spatial model is a reasonable approximation of the PLCP.

Dominant interference field. In wireless communication networks, due to the distance-dependent path-loss of the signal power, it is much more important to accurately capture the configuration of nodes in close proximity to the receiver of interest than the ones that are located farther away. For instance, among the tier 2 nodes, the nodes on the typical line are the set of dominant interferers due to their proximity to the typical receiver compared to the nodes located on the other lines. As we have retained the exact distribution of the tier 2 nodes on the typical line in the proposed approximate model, we can accurately characterize the dominant interference field. This will be clearly demonstrated through numerical results in the next subsection.

7.4.3   COVERAGE PROBABILITY

We will now compute the coverage probability for the vehicular network model in the presence of shadowing using the approximate model. We will first summarize the effective spatial model

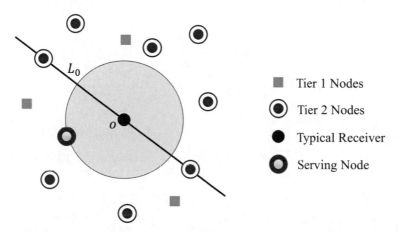

Figure 7.6: Illustration of the approximate spatial model.

after applying the approximation and accounting for the effect of shadowing. The tier 1 nodes are modeled by a homogeneous 2D PPP Φ_1' with density μ_1. The locations of tier 2 nodes are modeled by $\tilde{\Phi}_2$ which is the superposition of a 1D PPP Ψ_{L_0}' with density μ_2 along the line L_0 passing through the origin and a homogeneous 2D PPP Φ_a with density μ_a, as illustrated in Fig. 7.6. Note that this spatial model is quite similar to that of the canonical setup considered in Section 7.1 except for the additional 1D PPP Ψ_{L_0}' along the typical line. So, we will now present the coverage analysis for this setup by following the steps given in Fig. 7.1.

Note that the candidate serving nodes under the revised spatial model are simply the closest nodes to the typical receiver from tier 1 and tier 2. We denote the events in which the typical receiver connects to the tier 1 and tier 2 nodes by $\mathcal{E}_1$ and $\mathcal{E}_2$, respectively. Further, we denote the distance from the typical receiver to the candidate serving nodes from tier 1 and tier 2 by R_1 and R_2, respectively.

Step 1: Relevant distance distributions. The CDF and PDF of the distances R_1 and R_2 are given by

$$\text{CDF: } F_{R_1}(r_1) = 1 - \exp(-\mu_1 \pi r_1^2) \tag{7.72}$$
$$F_{R_2}(r_2) = 1 - \exp(-2\mu_2 r_2 - \mu_a \pi r_2^2), \tag{7.73}$$
$$\text{PDF: } f_{R_1}(r_1) = 2\pi \mu_1 r_1 \exp(-\mu_1 \pi r_1^2) \tag{7.74}$$
$$f_{R_2}(r_2) = (2\mu_2 + 2\mu_a \pi r_2) \exp(-2\mu_2 r_2 - \mu_a \pi r_2^2). \tag{7.75}$$

The distribution of R_1 directly follows from the nearest-neighbor distance distribution of a 2D PPP and the distribution of R_2 simply follows from the fact that the PPPs Ψ_{L_0}' and Φ_a are independent.

Step 2: Association probabilities. The probability of occurrence of the event $\mathcal{E}_1$ is computed as

$$\mathbb{P}(\mathcal{E}_1) = \mathbb{P}(P_1 B_1 R_1^{-\alpha} > P_2 B_2 R_2^{-\alpha}) = \int_0^\infty \left(1 - F_{R_2}\left(\frac{r_1}{\zeta}\right)\right) f_{R_1}(r_1) dr_1$$

$$= \int_0^\infty \exp\left(-\mu_1 \pi r_1^2 - 2\mu_2 \frac{r_1}{\zeta} - \mu_a \pi \frac{r_1^2}{\zeta^2}\right) 2\pi \mu_1 r_1 dr_1, \tag{7.76}$$

where $\zeta = \left(\frac{P_2 B_2}{P_1 B_1}\right)^{-\frac{1}{\alpha}}$. As the events $\mathcal{E}_1$ and $\mathcal{E}_2$ are complementary, $\mathbb{P}(\mathcal{E}_2) = 1 - \mathbb{P}(\mathcal{E}_1)$.

Step 3: Serving distance distribution. We will now determine the PDF of the serving distance R conditioned on the events $\mathcal{E}_1$ and $\mathcal{E}_2$. First, the CDF of R conditioned on $\mathcal{E}_1$ can be computed as

$$F_R(r \mid \mathcal{E}_1) = 1 - \mathbb{P}(R > r \mid \mathcal{E}_1) = 1 - \frac{1}{\mathbb{P}(\mathcal{E}_1)} \mathbb{P}(R_1 > r, R_1 < \zeta R_2)$$

$$= 1 - \frac{1}{\mathbb{P}(\mathcal{E}_1)} \int_{\frac{r}{\zeta}}^\infty \left(F_{R_1}(\zeta r_2) - F_{R_1}(r)\right) f_{R_2}(r_2) dr_2. \tag{7.77}$$

Substituting the expressions for $F_{R_1}(\cdot)$ and $f_{R_2}(r_2)$ in the above equation and taking the derivative using Leibniz's rule, we obtain the PDF of R conditioned on $\mathcal{E}_1$ as

$$f_R(r \mid \mathcal{E}_1) = \frac{1}{\mathbb{P}(\mathcal{E}_1)} 2\pi \mu_1 r \exp\left(-\mu_1 \pi r^2 - 2\mu_2 \frac{r}{\zeta} - \mu_a \pi \frac{r^2}{\zeta^2}\right). \tag{7.78}$$

Following the same procedure, we obtain the PDF of R conditioned on $\mathcal{E}_2$ as

$$f_R(r \mid \mathcal{E}_2) = \frac{1}{\mathbb{P}(\mathcal{E}_2)} (2\mu_2 + 2\mu_a \pi r) \exp(-2\mu_2 r - \mu_a \pi r^2 - \mu_1 \pi \zeta^2 r^2). \tag{7.79}$$

Step 4: Laplace transform of interference. We decompose the aggregate interference I at the typical receiver into two independent components I_1 and I_2 corresponding to the interference from the tier 1 and tier 2 nodes, respectively. We will first determine the Laplace transform of the distribution of interference conditioned on R and $\mathcal{E}_1$. This computation is quite similar to that of the canonical setup in Section 7.1. Conditioning on R and $\mathcal{E}_1$ implies that the closest interfering tier 1 node to the typical receiver is farther than R and the closest tier 2 node is farther than $\frac{R}{\zeta}$. Therefore, we have

$$\mathcal{L}_{I_1}(s \mid \mathcal{E}_1) = \mathbb{E}\left[\exp\left(-s \sum_{x \in \Phi_1' \setminus B(o,r)} P_1 H_1 \|x\|^{-\alpha}\right)\right]$$

$$= \exp\left(-2\pi \mu_1 \int_r^\infty \frac{s P_1 x^{1-\alpha}}{1 + s P_1 x^{-\alpha}} dx\right). \tag{7.80}$$

We will now calculate the Laplace transform of the distribution of interference from tier 2 nodes. As the spatial distribution of tier 2 nodes is the superposition of a 1D PPP along the typical line and a 2D PPP, we denote the interference from these two set of nodes by I_{20} and I_{21}, respectively, and we will compute them separately. First, the Laplace transform of the distribution of interference from tier 2 nodes on the typical line, denoted by I_{20}, can be computed as

$$
\mathcal{L}_{I_{20}}(s \mid \mathcal{E}_1) = \mathbb{E} \left[\exp \left(-s \sum_{x \in \Psi_L' \setminus B(o, \frac{r}{\xi})} P_2 H_2 \|x\|^{-\alpha} \right) \right]
$$
$$
= \exp \left(-2\mu_2 \int_{\frac{r}{\xi}}^{\infty} \frac{s P_2 x^{-\alpha}}{1 + s P_2 x^{-\alpha}} dx \right). \tag{7.81}
$$

Similarly, the Laplace transform of the distrubution of interference from the tier 2 nodes modeled by a 2D PPP is given by

$$
\mathcal{L}_{I_{21}}(s \mid \mathcal{E}_1) = \mathbb{E} \left[\exp \left(-s \sum_{x \in \Phi_a \setminus B(o, \frac{r}{\xi})} P_2 H_2 \|x\|^{-\alpha} \right) \right]
$$
$$
= \exp \left(-2\pi\mu_a \int_{\frac{r}{\xi}}^{\infty} \frac{s P_2 x^{1-\alpha}}{1 + s P_2 x^{-\alpha}} dx \right). \tag{7.82}
$$

As the interference components I_1, I_{20}, and I_{21} are independent, the Laplace transform of the distribution of the aggregate interference conditioned on R and $\mathcal{E}_1$ can simply be computed as the product of the Laplace transforms of the individual components of the interference as

$$
\mathcal{L}_I(s \mid \mathcal{E}_1) = \mathcal{L}_{I_1}(s \mid \mathcal{E}_1) \mathcal{L}_{I_{20}}(s \mid \mathcal{E}_1) \mathcal{L}_{I_{21}}(s \mid \mathcal{E}_1)
$$
$$
= \exp \left[-2\pi\mu_1 \int_r^{\infty} \frac{s P_1 x^{1-\alpha}}{1 + s P_1 x^{-\alpha}} dx - 2\mu_2 \int_{\frac{r}{\xi}}^{\infty} \frac{s P_2 x^{-\alpha}}{1 + s P_2 x^{-\alpha}} dx \right.
$$
$$
\left. - 2\pi\mu_a \int_{\frac{r}{\xi}}^{\infty} \frac{s P_2 x^{1-\alpha}}{1 + s P_2 x^{-\alpha}} dx \right]. \tag{7.83}
$$

Following the same procedure, the Laplace transform of the distribution of interference conditioned on R and $\mathcal{E}_2$ is obtained as

$$
\mathcal{L}_I(s \mid \mathcal{E}_2) = \exp \left[-2\pi\mu_1 \int_{\xi r}^{\infty} \frac{s P_1 x^{1-\alpha}}{1 + s P_1 x^{-\alpha}} dx - 2\mu_2 \int_r^{\infty} \frac{s P_2 x^{-\alpha}}{1 + s P_2 x^{-\alpha}} dx \right.
$$
$$
\left. - 2\pi\mu_a \int_r^{\infty} \frac{s P_2 x^{1-\alpha}}{1 + s P_2 x^{-\alpha}} dx \right]. \tag{7.84}
$$

Step 5: Coverage probability. For a desired SINR threshold β, the coverage probability conditioned on the events $\mathcal{E}_1$ and $\mathcal{E}_2$ can be computed by substituting the above results in (5.5).

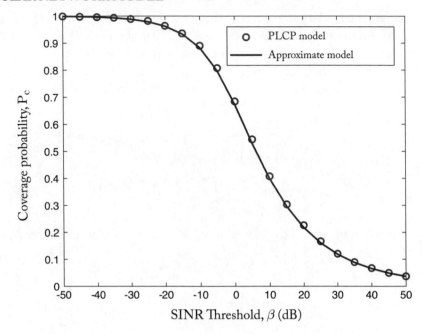

Figure 7.7: Coverage probability of the typical receiver as a function of SINR threshold ($\lambda_1 = 0.5$ nodes/km^2, $\mu_l = 10$ km^{-1}, $\lambda_2 = 5$ nodes/km, $P_1 = 43$ dBm, $P_2 = 23$ dBm, and $\alpha = 4$).

Further, using the law of total probability, the overall coverage probability is given by

$$
\begin{aligned}
\mathrm{P_c} = &\int_0^\infty \exp\left[-\frac{\beta r^\alpha}{P_1}\sigma^2 - 2\pi\mu_1 \int_r^\infty \frac{\beta r^\alpha x^{1-\alpha}}{1+\beta r^\alpha x^{-\alpha}}\mathrm{d}x - 2\mu_2 \int_{\frac{r}{\zeta}}^\infty \frac{\beta r^\alpha P_2 x^{-\alpha}}{P_1+\beta r^\alpha P_2 x^{-\alpha}}\mathrm{d}x \right.\\
&\left. - 2\pi\mu_a \int_{\frac{r}{\zeta}}^\infty \frac{\beta r^\alpha P_2 x^{1-\alpha}}{P_1+\beta r^\alpha P_2 x^{-\alpha}}\mathrm{d}x - \mu_1\pi r^2 - 2\mu_2\frac{r}{\zeta} - \mu_a\pi\frac{r^2}{\zeta^2} \right]2\pi\mu_1 r\,\mathrm{d}r\\
&+ \int_0^\infty \exp\left[-\frac{\beta r^\alpha}{P_2}\sigma^2 - 2\pi\mu_1 \int_{\zeta r}^\infty \frac{\beta r^\alpha P_1 x^{1-\alpha}}{P_2+\beta r^\alpha P_1 x^{-\alpha}}\mathrm{d}x - 2\mu_2 \int_r^\infty \frac{\beta r^\alpha x^{-\alpha}}{1+\beta r^\alpha x^{-\alpha}}\mathrm{d}x \right.\\
&\left. - 2\pi\mu_a \int_r^\infty \frac{\beta r^\alpha x^{1-\alpha}}{1+\beta r^\alpha x^{-\alpha}}\mathrm{d}x - 2\mu_2 r - \mu_a\pi r^2 - \mu_1\pi\zeta^2 r^2 \right]\\
&(2\mu_2 + 2\mu_a\pi r)\mathrm{d}r.
\end{aligned}
\tag{7.85}
$$

In Fig. 7.7, we compare the coverage probability of the typical receiver evaluated using the approximate model with that of the actual PLCP model and it can be observed that they match very closely, as expected. This further justifies that the proposed approximation of the PLCP is quite reasonable. This approximate model has also been employed in [85] to derive a closed-form expression for the success probability of the typical link in an ad hoc network.

7.5   SUMMARY

In this chapter, we presented the coverage probability of the typical vehicular user in a network where the vehicular nodes and RSUs coexist with cellular MBSs. Under the maximum average power based association policy, we first briefly summarized the key results involved in the computation of coverage probability for the canonical homogeneous 2D PPP model. We then provided a detailed derivation of the coverage probability for a setup where the vehicular nodes and RSUs are modeled by a PLCP. From the numerical results, we observed that the coverage probability improves as the density of vehicular nodes and RSUs increases and worsens as the line density increases. Finally, we discussed the analytical challenges involved in the computation of SINR in the presence of shadowing. Inspired by the asymptotic behavior of the PLCP, we presented a simple approximation of the PLCP model that lends tractability to the analysis without compromising the accuracy of our results.

CHAPTER 8

Load Analysis

In the previous chapter, we provided the SINR-based coverage analysis of the typical receiver in a vehicular communication network considering the cellular network model. It must be noted that the SINR measured at the typical receiver is agnostic to the distribution of the other receiver nodes in the network. However, in reality, the limited network resources are shared by all the users within the same cell. One of the key metrics that represents the demand on network resources is the number of users within the association region of a serving node, which is often referred to as the load on the serving node. Specifically, in a heterogeneous network, this information is quite useful in choosing appropriate values of selection bias in order to balance the load across different tiers and improve the network performance [86]. Therefore, in this chapter, we mainly focus on the characterization of load on MBSs and RSUs due to vehicular users which are modeled by a PLCP. The tools introduced in this chapter are different from the previous chapters and are hence useful in their own right. These results also enable the computation of the rate coverage of the typical receiver in the network [37, 70].

8.1 LOAD ON THE ROADSIDE UNITS

In this section, we will characterize the load on the RSUs due to vehicular users, where the locations of both these types of nodes are modeled by PLCPs on the same PLP. In order to clearly explain the challenges involved in the characterization of the load, we will begin our discussion with a simple single-tier network in which the vehicular users are served only by the RSUs. We will then present the load analysis for a more general case where vehicular users are served by both RSUs and MBSs in the network.

8.1.1 SINGLE-TIER NETWORK

In order to characterize the load on the typical RSU, we first need to identify its association region. Further, since the locations of vehicular users are coupled with the underlying PLP, we need to determine the distribution of the total length of line segments inside the association region of the typical RSU to compute the exact distribution of the load. In the single-tier network, the association regions of the RSUs are nothing but the cells of the Cox Voronoi tessellation as depicted in Fig. 8.1. Clearly, due to the randomness in the shape of the typical Cox Voronoi cell, it is quite hard to characterize the distribution of the total length of line segments inside the cell. Therefore, in the interest of analytical tractability, we assume that the RSUs serve only those vehicular users that are located on their own roads. This is not an unreasonable assumption

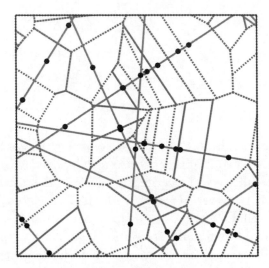

Figure 8.1: Illustration of the Cox Voronoi tessellation.

considering the fact that the information obtained from the nodes on the same road is much more relevant than the information obtained from the other roads. Using this assumption, we will first derive the load on the typical RSU in a vehicular network.

We consider a vehicular network in which the locations of vehicular users are modeled by a PLCP Φ_u such that the locations of vehicular nodes follow a homogeneous 1D PPP Ξ_L with density λ_u on each line L of the underlying PLP Φ_l with line density μ_l. We model the locations of RSUs by a PLCP Φ_2 on the same PLP Φ_l such that the density of RSUs distributed as a homogeneous 1D PPP Ψ_L on each line of Φ_l is λ_2. For this setup, the PMF of the load on the typical RSU is presented in the following lemma.

Lemma 8.1   *The PMF of the load on the typical RSU is*

$$\mathbb{P}(M_t = m) = \frac{4\lambda_2^2 \lambda_u^m (m+1)}{(\lambda_u + 2\lambda_2)^{m+2}}, \quad m = 0, 1, 2, \ldots. \tag{8.1}$$

Proof. Under the assumption that the RSUs serve only the vehicular users on their own road, the association regions of the RSUs on a line L are reduced to the Poisson Voronoi cells of the 1D PPP Ψ_L with density λ_2 along the line L. The exact distribution of the length of a 1D Poisson Voronoi cell is very well-known in the literature. Thus, the PDF of the length of the typical RSU cell is given by

$$f_{W_t}(w) = 4\lambda_2^2 w \exp(-2\lambda_2 w). \tag{8.2}$$

As the number of vehicular users on a line follows Poisson distribution, the PMF of the load on the typical RSU is computed as

$$\mathbb{P}(M_t = m) = \mathbb{E}_{W_t}\left[\mathbb{P}(M_t = m \mid W_t = w)\right]$$
$$= \int_0^\infty \frac{\exp(-\lambda_u w)(\lambda_u w)^m}{m!} 4\lambda_2^2 w \exp(-2\lambda_2 w)\mathrm{d}w. \tag{8.3}$$

Upon solving the above integral, we obtain the closed-form expression for the PMF of the load on the typical RSU. $\qquad\square$

We will now determine the PMF of the load on the RSU that serves the typical user, which is referred to as the *tagged RSU*. Owing to the stationarity of Φ_u, we assume that the typical user is located at the origin. Thus, the load on the tagged RSU is nothing but the load on the zero cell of the PVT corresponding to Ψ_L, which is given in the following lemma.

Lemma 8.2 *The PMF of the load on the tagged RSU is*

$$\mathbb{P}(M_z = m + 1) = \frac{4\lambda_2^3 \lambda_u^m (m+1)(m+2)}{(\lambda_u + 2\lambda_2)^{m+3}}, \quad m = 0, 1, 2, \ldots. \tag{8.4}$$

Proof. By length-biased sampling [87], the PDF of the length of the tagged RSU cell is obtained as

$$f_{W_z}(w) = \frac{w f_{W_t}(w)}{\int_0^\infty w f_{W_t}(w)\mathrm{d}w} = 4\lambda_2^3 w^2 \exp(-2\lambda_2 w). \tag{8.5}$$

Similar to Lemma 8.1, the PMF of the load on the tagged RSU can be computed as

$$\mathbb{P}(M_z = m + 1) = \mathbb{E}_{W_z}\left[\mathbb{P}(M_z = m \mid W_z = w)\right]$$
$$= \int_0^\infty \frac{\exp(-\lambda_u w)(\lambda_u w)^m}{m!} 4\lambda_2^3 w^2 \exp(-2\lambda_2 w)\mathrm{d}w. \tag{8.6}$$

We obtain the final expression for the PMF by solving the above integral. $\qquad\square$

8.1.2 MULTI-TIER NETWORK

Having derived the PMF of the load on the typical and tagged RSU in a single-tier vehicular network, we will now focus on the characterization of the load in a heterogeneous network where the vehicular users are served by both RSUs and cellular MBSs. We consider the same spatial model considered earlier in this section. However, additionally, we also model the locations of the MBSs by a homogeneous 2D PPP Φ_1 with density λ_1. As we have done previously, we will refer to the MBSs and RSUs as tier 1 and tier 2 nodes, respectively. We denote the transmit powers of tier 1 and tier 2 nodes by P_1 and P_2, respectively. We also assume that the vehicular users connect to the node that yields the maximum average received power. For this setup, under the assumption that the RSUs serve only those vehicular users on their own roads, we will characterize the PMF of the load on the typical and the tagged RSU.

We assume that the typical RSU is located at x_t on a line L. In this case, the boundary of the typical RSU cell is determined by the locations of the adjacent RSUs on the same line and also the MBSs in the network. We denote the typical RSU cell by $\mathcal{W}_{\text{typ}} = \{x \in L : P_2\|x - x_t\|^{-\alpha} \geq P_y\|x - y\|^{-\alpha}, \forall y \in \Phi_1 \cup \Psi_L\}$, where $P_y \in \{P_1, P_2\}$ denotes the transmit power of the node located at y. We will now present the PDF of the length of the typical RSU cell, denoted by $W_{\text{typ}} = \nu_1(\mathcal{W}_{\text{typ}})$, in the following lemma.

Lemma 8.3   *The PDF of the length of the typical RSU cell is*

$$
f_{W_{\text{typ}}}(w) \approx \int_{\frac{k+1}{2k}w}^{w} f_{W_1,1}(w - w_0) f_{W_0}(w_0) \mathrm{d}w_0 + \int_{\frac{k-1}{2k}w}^{\frac{k+1}{2k}w} f_{W_1,2}(w - w_0) f_{W_0}(w_0) \mathrm{d}w_0
$$
$$
+ \int_0^{\frac{k-1}{2k}w} f_{W_1,3}(w - w_0) f_{W_0}(w_0) \mathrm{d}w_0,
$$

(8.7)

where

$$
f_{W_1,1}(w_1) = 2\lambda_2 \exp(-2\lambda_2 w_1), \qquad 0 < w_1 < \frac{k-1}{k+1}w_0,
$$
$$
f_{W_1,2}(w_1) = \left[2\lambda_2 + \lambda_1 \frac{\partial \gamma_2(w_0, w_1)}{\partial w_1}\right] \exp\left[-2\lambda_2 w_1 - \lambda_1 \gamma_2(w_0, w_1)\right],
$$
$$
\frac{k-1}{k+1}w_0 < w_1 < \frac{k+1}{k-1}w_0,
$$
$$
f_{W_1,3}(w_1) = (2\lambda_2 + 2\lambda_1 \pi k^2 w_1) \exp\left[-2\lambda_2 w_1 - \lambda_1 \pi k^2 w_1^2 + \lambda_1 \pi k^2 w_0^2\right],
$$
$$
\frac{k+1}{k-1}w_0 < w_1 < \infty,
$$
$$
f_{W_0}(w_0) = (2\lambda_2 + 2\lambda_1 \pi k^2 w_0) \exp\left[-2\lambda_2 w_0 - \lambda_1 \pi k^2 w_0^2\right],
$$
$$
\gamma_2(w_0, w_1) = \pi k^2 w_1^2 - (kw_0)^2(\theta - \frac{1}{2}\sin(2\theta)) - (kw_1)^2(\phi - \frac{1}{2}\sin(2\phi)),
$$
$$
\theta = \arccos\left(\frac{(w_0 + w_1)^2 + (kw_0)^2 - (kw_1)^2}{2kw_0(w_0 + w_1)}\right),
$$
$$
\phi = \arccos\left(\frac{(w_0 + w_1)^2 - (kw_0)^2 + (kw_1)^2}{2kw_1(w_0 + w_1)}\right),
$$
$$
and\ k = \left(\frac{P_2}{P_1}\right)^{-\frac{1}{\alpha}}.
$$

Proof.   Let us denote the boundaries of the typical cell by s_0 and s_1 and the distances to these points from the typical RSU at x_t by W_0 and W_1, respectively. Thus, the length of the typical cell is given by $W_{\text{typ}} = W_0 + W_1$. Without loss of generality, we assume that s_0 is located to the right of x_t. We will begin with the computation of the marginal CDF of W_0. The CDF of W_0 can be computed as

$$F_{W_0}(w_0) = 1 - \mathbb{P}(W_0 > w_0).$$

The event $W_0 > w_0$ means that the distance to the nearest tier 2 node on the line L from the cell boundary s_0 is greater than w_0 and the distance to the nearest tier 1 node from s_0 is greater than $k w_0$, where $k = (\frac{P_2}{P_1})^{-\frac{1}{\alpha}}$. As the distribution of tier 1 and tier 2 nodes are independent, we obtain the CDF of W_0 as

$$\begin{aligned} F_{W_0}(w_0) &= 1 - \mathbb{P}\big(N_2(L \cap B(s_0, w_0)) = 0\big)\mathbb{P}\,(N_1\,(B\,(s_0, k w_0)) = 0) \\ &= 1 - \exp\left[-2\lambda_2 w_0 - \lambda_1 k^2 w_0^2\right], \end{aligned}$$

where $N_1(\cdot)$ and $N_2(\cdot)$ denote the number of tier 1 and tier 2 nodes, respectively, and $B(c, d)$ denotes a ball of radius d centered at c.

Recall that the cell boundary s_0 is either determined by the adjacent tier 2 node located on the same line or a tier 1 node. We will now discuss the different conditions imposed by these two cases in the computation of the CDF of W_1 conditioned on W_0.

First, let us consider the case where the cell boundary s_0 is determined by an adjacent tier 2 node on the same line. We will denote this event by $\mathcal{H}_1$. From the derivation of the marginal CDF of W_0, we already know that conditioning on W_0 implies the occurrence of the events $N_2(L \cap B(s_0, w_0)) = 0$ and $N_1(B(s_0, k w_0)) = 0$. Further, in the event of $\mathcal{H}_1$, conditioning on W_0 also means that there exists a tier 2 node on the line exactly at a distance w_0 to the right of the point s_0 (at a distance $2w_0$ from x_t) on the line L. However, the distribution of tier 2 nodes to the left of x_t on the line L is independent of the distribution of the tier 2 nodes to the right of x_t. Therefore, in the event of $\mathcal{H}_1$, the only condition that needs to be handled in the computation of the CDF of W_1 conditioned on W_0 is $N_1(B(s_0, k w_0)) = 0$.

We will now consider the case where the cell boundary s_0 is determined by a tier 1 node and we denote this event by $\mathcal{H}_2$. In this case, conditioning on W_0 additionally imposes the condition that there exists a tier 1 node on the circumference of the disc $B(s_0, k w_0)$. Thus, in the event of $\mathcal{H}_2$, conditioning on W_0 implies $N_2(L \cap B(s_0, w_0)) = 0$, $N_1(B(s_0, k w_0)) = 0$ and $N_1(\partial B(s_0, k w_0)) = 1$, where $\partial B(s_0, k w_0)$ denotes the boundary of the disc. Again, as explained earlier, the distribution of tier 2 nodes to the left of x_t is independent of the distribution of tier 2 nodes to the right of x_t. Therefore, in order to compute the conditional CDF $F_{W_1}(w_1|w_0)$, we need to include the two cases, $N_1(B(s_0, k w_0)) = 0$ and $N_1(\partial B(s_0, k w_0)) = 1$.

Thus, the CDF of W_1 conditioned on W_0 is computed as

$$\begin{aligned} F_{W_1}(w_1|w_0) &= \mathbb{P}(\mathcal{H}_1)\mathbb{P}(W_1 < w_1|W_0, \mathcal{H}_1) + \mathbb{P}(\mathcal{H}_2)\mathbb{P}(W_1 < w_1|W_0, \mathcal{H}_2) \\ &= \mathbb{P}(\mathcal{H}_1)\mathbb{P}\,(W_1 < w_1|N_1\,(B\,(s_0, k w_0)) = 0) \\ &\quad + \mathbb{P}(\mathcal{H}_2)\mathbb{P}\,(W_1 < w_1|N_1\,(B\,(s_0, k w_0)) = 0, N_1\,(\partial B\,(s_0, k w_0)) = 1)\,. \end{aligned}$$

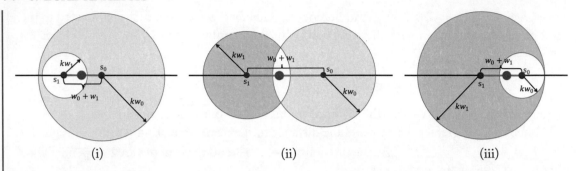

Figure 8.2: An illustration of the three cases: (i) $0 < w_1 < \frac{k-1}{k+1}w_0$, (ii) $\frac{k-1}{k+1}w_0 < w_1 < \frac{k+1}{k-1}w_0$, and (iii) $\frac{k+1}{k-1}w_0 < w_1 < \infty$.

Hence, in order to determine the exact conditional CDF $F_{W_1}(w_1|w_0)$, we need to first compute the probability of occurrence of the events $\mathcal{H}_1$ and $\mathcal{H}_2$. Further, the computation of other terms in the above equation is quite complex as we need to consider multiple cases depending on the relative ranges of w_1 and w_0. Therefore, in the interest of tractability, we do not distinguish these two cases in our analysis and ignore the additional conditions resulting from conditioning on the events $\mathcal{H}_1$ and $\mathcal{H}_2$. So, we derive the approximate CDF $F_{W_1}(w_1|w_0)$ by simply conditioning on the event $N_1(B(s_0, kw_0)) = 0$. Thus, the CDF of W_1 conditioned on W_0 can be computed as

$$
\begin{aligned}
F_{W_1}&(w_1|w_0) \\
&\approx 1 - \mathbb{P}\Big(\{N_2(L \cap B(s_1, w_1)) = 0\} \cap \{N_1(B(s_1, kw_1)) = 0\}\Big| N_1(B(s_0, kw_0)) = 0\Big) \\
&= 1 - \mathbb{P}\big(N_2(L \cap B(s_1, w_1)) = 0\big)\mathbb{P}\big(N_1(B(s_1, kw_1) \setminus B(s_0, kw_0)) = 0\big) \\
&= 1 - \exp(-2\lambda_2 w_1)\exp\big(-\lambda_1 \nu_2(B(s_1, kw_1) \setminus B(s_0, kw_0))\big).
\end{aligned}
\tag{8.8}
$$

We need to consider three different cases based on the relative ranges of w_0 and w_1 as illustrated in Fig. 8.2. Hence, the area of the region $B(s_1, kw_1) \setminus B(s_0, kw_0)$ is a piecewise function which is given by

$$
\nu_2\Big(B(s_1, kw_1) \setminus B(s_0, kw_0)\Big) = \begin{cases} \gamma_1(w_0, w_1), & 0 < w_1 < \dfrac{k-1}{k+1}w_0, \\[2mm] \gamma_2(w_0, w_1), & \dfrac{k-1}{k+1}w_0 < w_1 < \dfrac{k+1}{k-1}w_0, \\[2mm] \gamma_3(w_0, w_1), & \dfrac{k+1}{k-1}w_0 < w_1 < \infty \end{cases}
$$

$$= \begin{cases} 0, & 0 < w_1 < \dfrac{k-1}{k+1} w_0, \\[2ex] \pi k^2 w_1^2 - (kw_0)^2 \left(\theta - \dfrac{1}{2} \sin(2\theta) \right) \\ \qquad - (kw_1)^2 \left(\phi - \dfrac{1}{2} \sin(2\phi) \right), & \dfrac{k-1}{k+1} w_0 < w_1 < \dfrac{k+1}{k-1} w_0 \\[2ex] \pi (kw_1)^2 - \pi (kw_0)^2, & \dfrac{k+1}{k-1} w_0 < w_1 < \infty, \end{cases} \quad (8.9)$$

where

$$\theta = \arccos\left(\frac{(w_0 + w_1)^2 + (kw_0)^2 - (kw_1)^2}{2kw_0(w_0 + w_1)} \right),$$

$$\text{and } \phi = \arccos\left(\frac{(w_0 + w_1)^2 - (kw_0)^2 + (kw_1)^2}{2kw_1(w_0 + w_1)} \right).$$

Substituting (8.9) in (8.8), we obtain the conditional CDF of W_1 as follows:

$$F_{W_1}(w) = \begin{cases} 1 - e^{-2\lambda_2 w_1}, & 0 < w_1 < \dfrac{k-1}{k+1} w_0, \\[1.5ex] 1 - e^{-2\lambda_2 w_1 - \lambda_1 \gamma_2(w_0, w_1)}, & \dfrac{k-1}{k+1} w_0 < w_1 < \dfrac{k+1}{k-1} w_0, \\[1.5ex] 1 - e^{-2\lambda_2 w_1 - \lambda_1 \pi k^2 w_1^2 + \lambda_1 \pi k^2 w_0^2}, & \dfrac{k+1}{k-1} w_0 < w_1 < \infty. \end{cases} \quad (8.10)$$

Thus, the CDF of the length of the typical cell can be computed as

$$F_{W_{\text{typ}}}(w) = \mathbb{P}(W_0 + W_1 < w) = \int_0^w \mathbb{P}(W_1 < w - w_0 | W_0) f_{W_0}(w_0) \mathrm{d}w_0$$

$$= \int_0^w F_{W_1}(w - w_0 | w_0) f_{W_0}(w_0) \mathrm{d}w_0$$

$$= \int_{\frac{k+1}{2k} w}^w F_{W_1,1}(w - w_0) f_{W_0}(w_0) \mathrm{d}w_0 + \int_{\frac{k-1}{2k} w}^{\frac{k+1}{2k} w} F_{W_1,2}(w - w_0) f_{W_0}(w_0) \mathrm{d}w_0$$

$$+ \int_0^{\frac{k-1}{2k} w} F_{W_1,3}(w - w_0) f_{W_0}(w_0) \mathrm{d}w_0, \quad (8.11)$$

where $F_{W_1,1}(\cdot)$, $F_{W_1,2}(\cdot)$, and $F_{W_1,3}(\cdot)$ correspond to the expressions given in the piecewise CDF in (8.10). We obtain the PDF of W_{typ} by taking the derivative of $F_{W_{\text{typ}}}(w)$ w.r.t. w. $\qquad \square$

Using this result, the PMF of the load on the typical and tagged RSUs are presented in the following lemmas.

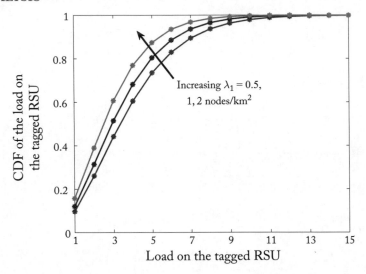

Figure 8.3: CDF of the load on the tagged RSU in a heterogeneous network ($\mu_l = 10$ km^{-1}, $\lambda_2 = 5$ nodes/km, $\lambda_u = 15$ nodes/km, $P_1 = 43$ dBm, $P_2 = 23$ dBm, and $\alpha = 4$).

Lemma 8.4 *The PMF of the load on the typical RSU is*

$$\mathbb{P}(M_t = m | \mathcal{E}_2) = \int_0^\infty \frac{\exp(-\lambda_u w)(\lambda_u w)^m}{m!} f_{W_{\text{typ}}}(w) \mathrm{d}w, \qquad m = 0, 1, 2, \ldots, \qquad (8.12)$$

where $f_{W_{\text{typ}}}(w)$ is given in Lemma 8.3

Proof. This result simply follows from the Poisson distribution of the vehicular users in the typical cell of the RSU. □

Lemma 8.5 *The PMF of the load on the tagged RSU is*

$$\mathbb{P}(M_z = m + 1) = \int_0^\infty \frac{\exp(-\lambda_u w)(\lambda_u w)^m}{m!} f_{W_{\text{tag}}}(w) \mathrm{d}w, \qquad m = 0, 1, 2, \ldots, \qquad (8.13)$$

where

$$f_{W_{\text{tag}}}(w) = \frac{w f_{W_{\text{typ}}}(w)}{\int_0^\infty w f_{W_{\text{typ}}}(w) \mathrm{d}w}. \qquad (8.14)$$

Proof. Similar to Lemma 8.2, this result follows from the Poisson distribution of vehicular users and length-biased sampling. □

In Fig. 8.3, we plot the CDF of the load on the tagged RSU for different values of the density of tier 1 nodes (MBSs) in the network. As expected, we observe that the load on the RSUs decreases as the density of MBSs increases.

8.2 LOAD ON THE MACRO BASE STATIONS

In this section, we will characterize the load on cellular MBSs due to vehicular users whose locations are modeled by PLCP. Similar to the previous section, we will first characterize the load on MBSs in a single-tier network in which the vehicular users are served only by the MBSs and then discuss a more general heterogeneous network model.

8.2.1 SINGLE-TIER NETWORK

We consider a vehicular network in which the locations of cellular MBSs are modeled by a homogeneous 2D PPP Φ_1 with density λ_1. We model the locations of vehicular users by a PLCP Φ_u such that the locations of vehicles on each line of the underlying PLP Φ_l follows a 1D PPP with intensity λ_u and the line density of the PLP is μ_l. Further, we assume that all the MBSs have the same transmit power. We assume that each vehicular user connects to the MBS that yields the maximum average received power. Under this policy, the association regions of the MBSs are nothing but the cells of the PVT corresponding to the PPP Φ_1. For this network, we will first focus on the computation of the load on the typical cellular MBS.

Recall that the PMF of the number of points of the PLCP in the typical cell of a PVT has already been computed in Lemma 4.6. Upon substituting λ_u and λ_1 in place of λ_p and η, respectively, in (4.11), we obtain the exact PMF of the load on the typical MBS. Although the exact expression for the PMF of the load clearly captures the dependence of the load on other sources of randomness, such as the perimeter of the cell, it does not lend itself well to numerical evaluation. Therefore, we propose an accurate approximation for the PMF of the load on the typical MBS based on the following assumption.

Assumption 8.6 The distribution of the number of vehicular users in the typical Poisson Voronoi cell is the same as that of the number of vehicular users in the disc $B(o, R_t)$ whose area is equal to that of the typical cell, i.e., $\pi R_t^2 = |V_o|$.

This assumption is primarily inspired by the well-known asymptotic result in the stochastic geometry literature that large Poisson Voronoi cells are circular [88]. As we will demonstrate shortly, the PMF of the load computed using this approximation is quite accurate in the non-asymptotic regime as well. Using this assumption, we will now derive the approximate PMF of the load on the typical MBS in the following lemma.

Lemma 8.7 *Under Assumption 8.6, the PMF of the load on the typical MBS is*

$$\mathbb{P}(M_t = m) \approx \frac{(-\lambda_u)^m}{m!}\left[\frac{\partial^m}{\partial s^m}\int_0^\infty \exp\left[-2\pi\lambda_l r_t + 2\pi\lambda_l \int_0^{r_t} e^{-2s\sqrt{r_t^2-\rho^2}}\mathrm{d}\rho\right]f_{R_t}(r_t)\mathrm{d}r_t\right]_{s=\lambda_u},$$
$$m = 0, 1, 2, \ldots, \qquad (8.15)$$

where $f_{R_t}(r_t) = 2\pi r_t \frac{ab^{\frac{c}{a}}}{\Gamma(\frac{c}{a})}\lambda_1^c(\pi r_t^2)^{c-1}\exp(-b(\lambda_1\pi r_t^2)^a),$ $\qquad (8.16)$

$a = 1.07950,\ b = 3.03226,\ and\ c = 3.31122.$

Proof. We now need to compute the PMF of the number of users in Φ_u in the disc $B(o, R_t)$. Since the area of the typical cell is random, the radius R_t here is also random. The PMF of the number of points of the PLCP in a disc of fixed radius is already derived in Lemma 4.4. Thus, upon conditioning on the radius of the disc R_t, the PMF of the load on the typical MBS is given by

$$\mathbb{P}(M_t = m \mid R_t) \approx \frac{(-\lambda_u)^m}{m!}\left[\frac{\partial^m}{\partial s^m}\mathcal{L}_T(s|R_t)\right]_{s=\lambda_u}, \qquad (8.17)$$

where

$$\mathcal{L}_T(s|R_t) = \exp\left[-2\pi\lambda_l\int_0^{r_t} 1 - \exp(-2s\sqrt{r_t^2-\rho^2})\mathrm{d}\rho\right]. \qquad (8.18)$$

From [78], we know that the PDF of the area of the typical Poisson Voronoi cell is

$$f_{|V_o|}(z) = \frac{ab^{\frac{c}{a}}}{\Gamma(\frac{c}{a})}\lambda_1^c z^{c-1}\exp(-b(\lambda_1 z)^a), \qquad (8.19)$$

where $a = 1.07950$, $b = 3.03226$, and $c = 3.31122$. Since $R_t = \sqrt{\frac{|V_o|}{\pi}}$, the PDF of R_t is given by $f_{R_t}(r_t) = 2\pi r_t f_{|V_o|}(\pi r_t^2)$. Taking the expectation of (8.17) w.r.t. R_t, we obtain the final expression. $\qquad \square$

We will refer to the serving MBS of the typical user as the *tagged MBS* and the corresponding cell as the *tagged cell*. Without loss of generality, we assume that the typical user is located at the origin. Hence, the tagged cell is the same as the zero cell $V(o)$. We will now derive the PMF of the load on the tagged MBS in the following lemma by using an assumption similar to that of Assumption 8.6.

Lemma 8.8 *The PMF of the load on the tagged MBS is*

$$\mathbb{P}(M_z = m) \approx \frac{(-\lambda_u)^m}{m!} \int_0^\infty \int_0^\infty \left[\frac{\partial^m}{\partial s^m} \exp\left[-st_0 - 2\pi\lambda_l \int_0^{r_z} 1 - e^{-2s\sqrt{r_z^2 - \rho^2}} \, d\rho \right] \right]_{s=\lambda_u}$$
$$\times f_{R_z}(r_z) f_{T_0}(t_0) dt_0 dr_z, \quad m = 0, 1, 2, \ldots, \quad (8.20)$$

where $f_{T_0}(t_0) = \dfrac{4\sqrt{\lambda_1}}{\pi} t_0 f_C(t_0)$, $\quad f_{R_z}(r_z) = 2\lambda_1\pi^2 r_z^3 \dfrac{ab^{\frac{c}{a}}}{\Gamma(\frac{c}{a})} \lambda_1^c (\pi r_z^2)^{c-1} \exp(-b(\lambda_1\pi r_z^2)^a)$,

$a = 1.07950$, $b = 3.03226$, *and* $c = 3.31122$.

Proof. Recall that under the Palm probability, the point process of vehicular users is the superposition of the PLCP Φ_u, a 1D PPP Ξ_{L_0} with intensity λ_u, and an atom at the origin. In order to compute the load on the tagged MBS, we first need to characterize the total chord length T of the lines of the PLP Φ_l in the zero cell. Clearly, the typical line L_0 always intersects the zero cell and there are also a random number of lines that intersect the zero cell. So, we decompose the total chord length T into two components: (i) the length of the chord T_0 corresponding to L_0, and (ii) total length of remaining chords T_1 excluding the typical line. We will determine the Laplace transforms of the distributions of T_0 and T_1 separately.

First, we will focus on T_0. Recall that in Section 2.3.3, we considered the point process formed by the intersection of the PVT with an arbitrary line and derived its intensity in Lemma 2.13. Without loss of generality, let us assume this line to be aligned along the x-axis. We denote the 1D point process formed by the intersection points along the x-axis by Ψ_x. In that setting, the length of an arbitrarily chosen interval (inter-point distance) of Ψ_x is the same as the chord length of the line intersecting the typical cell, whose PDF $f_C(c)$ is given in Lemma 2.14. Now, let us focus on the interval that contains the origin, which will be referred to as the zero interval. From [87], it follows that the PDF of the length of the zero interval is the length-biased PDF of the chord length in the typical cell. Note that the length of this zero interval is the same as the chord length of the typical line L_0 in the zero cell of the PVT. Therefore, the PDF of T_0 is given by

$$f_{T_0}(t_0) = \frac{t_0 f_C(t_0)}{\mathbb{E}[C]}. \quad (8.21)$$

Thus, we obtain

$$\mathcal{L}_{T_0}(s) = \int_0^\infty \exp(-st_0) f_{T_0}(t_0) dt_0. \quad (8.22)$$

We will now compute the Laplace transform of the distribution of T_1. Similar to the previous lemma, we now assume that the distribution of number of vehicular users in the zero cell on the lines excluding the typical line is the same as that of the number of users in a disc of radius R_z whose area is equal to that of the zero cell, i.e., $\pi R_z^2 = |V(o)|$. Following the same procedure as in the previous lemma, upon conditioning on the radius R_z, we obtain the Laplace transform of the distribution of T_1 as

$$\mathcal{L}_{T_1}(s \mid R_z) \approx \exp\left[-2\pi\lambda_l \int_0^{r_z} 1 - e^{-2s\sqrt{r_z^2 - \rho^2}} \mathrm{d}\rho\right]. \tag{8.23}$$

By area-biased sampling [87], the PDF of the area of the zero cell is

$$f_{|V(o)|}(z) = \frac{z f_{|V_o|}(z)}{\mathbb{E}[|V_o|]} = \lambda_1 z f_{|V_o|}(z).$$

Since $R_z = \sqrt{\frac{|V(o)|}{\pi}}$, the PDF of R_z is

$$f_{R_z}(r_z) = 2\lambda_1 \pi^2 r_z^3 f_{|V_o|}(\pi r_z^2).$$

We can now determine the Laplace transform of the distribution of T_1 by computing the expectation of the expression in (8.23) w.r.t. R_z. As T_0 and T_1 are independent, the Laplace transform of the overall chord length distribution can be computed as

$$\mathcal{L}_T(s) = \mathcal{L}_{T_0}(s)\mathcal{L}_{T_1}(s). \tag{8.24}$$

From Lemma 4.4, we know that the PMF of the number of users in the zero cell can be expressed in terms of the Laplace transform of the chord length distribution as

$$\mathbb{P}(M_z = m + 1) = \frac{(-\lambda_u)^m}{m!}\left[\frac{\partial^m}{\partial s^m}\mathcal{L}_T(s)\right]_{s=\lambda_u}. \tag{8.25}$$

Substituting the expression for $\mathcal{L}_T(s)$ in the above equation, we obtain the final expression for the PMF of the load on the tagged MBS. $\square$

8.2.2   MULTI-TIER NETWORK

We will now characterize the load on the MBSs in a heterogeneous network where vehicular users are served by both RSUs and MBSs. In addition to the spatial model that we have considered earlier in this section, we also model the locations of the RSUs by a PLCP Φ_2 such that the locations of the RSUs on each line of the PLP Φ_l follows a homogeneous 1D PPP with density λ_2. As we have done in the previous section, in the interest of analytical tractability, we assume that the RSUs serve only those vehicular users on their own roads. Further, we assume that the vehicular users connect to the node that yields the maximum average received power.

As shown in the previous subsection, we need to characterize the total length of the lines that are covered by the MBS in order to compute the PMF of the load on the MBS. While we

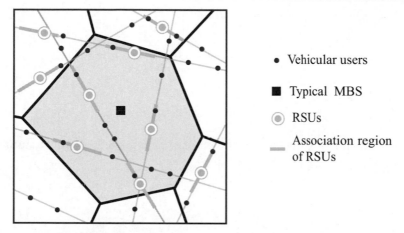

Figure 8.4: Illustration of the typical MBS cell in a heterogeneous network.

were able to characterize the total length of lines in the association region of an MBS in a single-tier network, it is much more challenging in a multi tier network. Although the association regions of the MBSs are still the Poisson Voronoi cells formed by the PPP Φ_1, some of the vehicular users within these cells are served by their nearest RSUs on their own lines as depicted in Fig. 8.4. Therefore, in order to compute the exact PMF of the load on the MBSs, we need to determine the PDF of the length of the chord segments within the MBS cell that are not covered by RSUs, which is not feasible. Therefore, in this subsection, we will only focus on the computation of the first moment of the load on MBSs which has been recently studied in [37, 89].

We will first focus on the load on the typical MBS. As we had mentioned, some of the vehicular users in the typical MBS cell are served by the RSUs located in the cell. So, we can compute the mean load on the typical MBS by subtracting the mean number of users served by RSUs within the typical cell from the mean of total number of vehicular users in the typical cell, which is presented in the following lemmas.

Lemma 8.9 *The mean of the total number of vehicular users in the typical MBS cell is $\frac{\mu_l \lambda_u}{\lambda_1}$.*

Proof. From Lemma 2.10, it follows that the mean area of the typical cell of a cellular MBS is given by $\frac{1}{\lambda_1}$. As the mean line length per unit area of the PLP Φ_l is μ_l and the mean number of points per unit length on each line is λ_u, the mean number of users in the typical MBS cell is simply given by $\frac{\mu_l \lambda_u}{\lambda_1}$. □

Lemma 8.10   *The mean load on the typical MBS is*

$$\frac{\mu_l \lambda_u}{\lambda_1} - \frac{\mu_l \lambda_2}{\lambda_1} \lambda_u \int_0^\infty w f_{W_{\text{typ}}}(w) \mathrm{d}w, \tag{8.26}$$

where $f_{W_{\text{typ}}}(w)$ is given in Lemma 8.3.

Proof. Following the same argument as in the proof of Lemma 8.9, the mean number of RSUs inside the typical MBS cell is given by

$$\frac{\mu_l \lambda_2}{\lambda_1}. \tag{8.27}$$

In Lemma 8.3, we have already derived the PDF of the length of the typical RSU cell. Therefore, using this result, the mean number of vehicular users inside the typical MBS cell that are served by RSUs can be computed as

$$\frac{\mu_l \lambda_2}{\lambda_1} \lambda_u \int_0^\infty w f_{W_{\text{typ}}}(w) \mathrm{d}w. \tag{8.28}$$

Thus, the mean load on the typical MBS can be obtained by subtracting the mean number of vehicular users served by RSUs inside the typical MBS cell, which is given by the above expression, from the mean number of vehicular users within the typical MBS cell.   □

We will now derive the mean load on the tagged cellular MBS in the multi-tier vehicular network. As done in the case of the typical MBS, we first determine the mean of the total number of vehicular users in the tagged MBS cell.

Lemma 8.11   *The mean of the total number of vehicular users in the tagged MBS cell is $1 + \frac{1.28 \mu_l \lambda_u}{\lambda_1} + \frac{3.216}{\pi \sqrt{\lambda_1}}$.*

Proof. Recall that under the Palm probability, the point process of vehicular users is the superposition of the PLCP Φ_u, a 1D PPP Ξ_{L_0} with intensity λ_u, and an atom at the origin. We will calculate the mean number of users inside the tagged cell from each of these components separately. The mean number of users from Φ_u that lie inside the tagged cell is given by

$$\mathbb{E}[N_u(\Phi_u \cap V(o))] = \lambda_u \mu_l |V(o)| = \frac{1.28 \mu_l \lambda_r}{\lambda_1}. \tag{8.29}$$

We will now compute the mean number of users on the typical line in the tagged cell by first determining the mean length of the chord segment corresponding to L_0 in the tagged cell. In the proof of Lemma 8.8, we have already shown that the PDF of the length of chord segment of the typical line in the zero cell is given by

$$f_{T_0} = \frac{4\sqrt{\lambda_1}}{\pi} t_0 f_C(t_0). \tag{8.30}$$

Therefore, the mean length of the chord segment of the typical line in the tagged cell can be simply computed as

$$\int_0^\infty t_0 f_{T_0}(t_0) \mathrm{d}t_0 = \frac{3.216}{\pi\sqrt{\lambda_1}}. \tag{8.31}$$

Thus, the mean number of users on the typical line, including the typical user at the origin, is given by

$$\mathbb{E}[N_u(\{\Psi_u \cup \{o\}\} \cap V(o))] = 1 + \lambda_u \frac{3.216}{\pi\sqrt{\lambda_1}}. \tag{8.32}$$

Combining (8.29) and (8.32), we obtain the final expression for the mean number of users in the tagged MBS cell. $\qquad \square$

Lemma 8.12 *The mean load on the tagged MBS is*

$$1 + \lambda_u \left(\frac{1.28\mu_l}{\lambda_1} + \frac{3.216}{\pi\sqrt{\lambda_1}} \right) \left(1 - \lambda_2 \int_0^\infty w f_{W_{\mathrm{typ}}}(w) \mathrm{d}w \right). \tag{8.33}$$

Proof. Similar to Lemma 8.10, this result can be obtained by subtracting the mean number of vehicular users that are served by RSUs from the mean number of vehicular users inside the cell. $\qquad \square$

8.3 SUMMARY

In this chapter, we characterized the load on the cellular MBSs and RSUs due to vehicular users whose locations are modeled by a PLCP. First, we presented the PMF of the load on the typical and the tagged RSUs in a single-tier network where vehicular users are served only by RSUs located on their own roads. We then considered a heterogeneous network where RSUs coexist with MBSs and computed the PMF of the load by deriving the expression for the PDF of the length of the typical RSU cell. In the same manner, we computed the PMF of the load on the typical and the tagged MBS in a single-tier vehicular network and then calculated the mean load on the typical and the tagged cellular MBS for the heterogeneous network model.

Thus far, in Chapters 5–8, we have exclusively focused on the applications of PLP and PLCP to vehicular communication networks. However, given the ability of this modeling approach to capture the physical layout of the roads through a stochastic model, it can be applied to many other areas as well. Two concrete examples of applications beyond vehicular communication networks are provided in the next two chapters.

CHAPTER 9

Localization Networks

Accurate localization is critical to several indoor and outdoor applications such as autonomous driving, asset tracking, sensor-data collection, and search-and-rescue operations. In many of these applications, the desired accuracy is of the order of a few centimeters, which is not always possible through the global positioning system (GPS). While GPS is usually quite accurate in outdoor clear sky conditions, it may not be very reliable when the receiver does not have a clear LoS to the minimum required number of satellites, such as in urban canyons and indoors. In such scenarios, the fallback option is to use terrestrial wireless networks to obtain a position fix. One way of obtaining a position fix in 2D is by estimating the distance of the target from three anchor points (e.g., cellular BSs) whose positions are known. One of the ranging techniques for estimating this distance is to use the time-of-arrival (ToA) information of a ranging signal along the LoS path. However, in practice, the LoS link between an anchor and the target may be blocked by obstacles in the environment. These obstacles could be static, such as buildings and trees, or mobile such as vehicles. In 2D, a target is said to be in a blind spot if it is in the LoS path of less than three anchors. Owing to the randomness of the locations of anchors and the obstacles, there is a probability associated with the typical target being in blind spot, which is studied in this chapter.

There have been a few works in the literature where the effect of blocking due to obstacles has been studied using tools from random shape theory [90, 91]. In [91], the buildings in an urban area were modeled as a process of rectangles with random areas and the centers of the rectangles form a 2D PPP. However, since the blockage of the LoS path of the signal from the target to the anchors is agnostic to the thickness of the obstacles, it is sufficient to model the obstacles as line segments. Therefore, we employ a Boolean model to model these line segments, where the mid-points of the line segments are distributed as a 2D PPP and the length of the line segments are random. It is quite evident that the blocking correlation increases as the length of the line segments increases. In this book, we will demonstrate the use of PLP to model the worst-case scenario of correlated blocking and compute the asymptotic blind spot probability of the typical target node. More details about the system model are provided next.

9.1 SYSTEM MODEL

We model the location of anchor nodes in the network by a homogeneous 2D PPP Φ_c with density λ. Owing to the stationarity of Φ_c, we assume that the typical target node is located at the origin. We model the obstacles by a PLP Φ_l such that the projection of the origin onto the

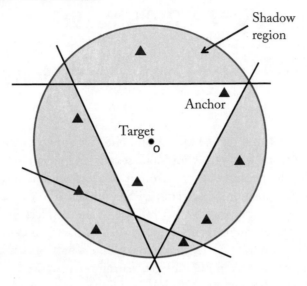

Figure 9.1: Illustration of the system model.

lines of Φ_l forms a homogeneous 2D PPP Φ_b in $\mathbb{R}^2$ with density λ_0, as illustrated in Fig. 9.1. We assume a ToA-based localization where the range of the target is estimated at each anchor using the ranging signal transmitted omni-directionally from the target. The transmit power limitation induces the constraint that the anchor nodes must lie within a disc $B(o, d)$ of radius d centered at the origin for the target to be localized.

9.2   BLIND SPOT PROBABILITY

As discussed earlier, the probability of a blind spot depends on the distribution of the visible anchors in the network. The anchors that are behind the obstacles as seen from the target are said to be in the shadow region and the rest of the region inside the disc $B(o, d)$ is called the visible region, as illustrated in Fig. 9.1. We denote the visible region of the typical target, which is nothing but the polygon formed by the lines of the PLP enclosing the origin, by C_o and its area by A_v. Further, we are only interested in the case where $C_o \subseteq B(o, d)$ as the blind spot probability depends only on the area of C_o, which is analytically tractable. Conditioned on the visible region C_o, the blind spot probability of the typical target is computed as

$$
\begin{aligned}
P_{b|C_o} &= \mathbb{P}(N_p(\Phi_c \cap C_o) < 3) \\
&= \exp(-\lambda_c a_v)\left(1 + \lambda_c a_v + \frac{(\lambda_c a_v)^2}{2}\right).
\end{aligned} \tag{9.1}
$$

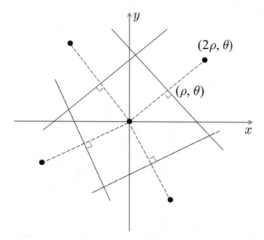

Figure 9.2: Illustration of the polygon formed by PLP enclosing the origin as a Poisson Voronoi cell.

Therefore, by taking the expectation of the above expression w.r.t. A_v, the overall blind spot probability can be computed as

$$P_b = \mathbb{E}_{A_v}\left[e^{-\lambda_c a_v}\left(1 + \lambda_c a_v + \frac{(\lambda_c a_v)^2}{2}\right)\right]. \tag{9.2}$$

We now need to compute the distribution of the area of the visible region C_o. The area of this region can be computed using its relation to a Poisson-Voronoi cell as given in the following theorem [62].

Theorem 9.1   The distribution of the area of the visible region C_o coincides with that of the typical Poisson Voronoi cell with density $\lambda_0/4$.

Proof. By construction, a line represented by polar coordinates (ρ, θ) is the perpendicular bisector of the line joining the origin and the point with coordinates $(2\rho, \theta)$, as shown in Fig. 9.2. The set of points $2\Phi_b$ is also a 2D PPP with density $\lambda_0/4$. By Slivnyak's theorem, the addition of a point at the origin to the 2D PPP $2\Phi_b$ does not change its distribution. Thus, the polygon C_o formed by the lines of the PLP can be interpreted as the typical Poisson Voronoi cell of the 2D PPP $2\Phi_b$ with density $\lambda_0/4$. $\square$

From [78, 92], we already know that the PDF of the area of the cell A_v is given by

$$f_{A_v}(a_v) = \frac{ab^{(c/a)}}{\Gamma(c/a)}\left(\frac{\lambda_0}{4}\right)^c a_v^{c-1}e^{-b(\lambda_0 a_v/4)^a}, \tag{9.3}$$

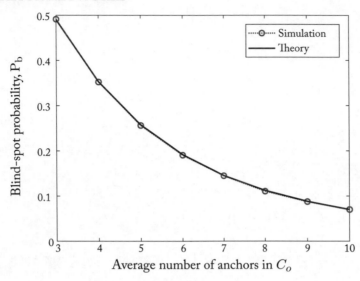

Figure 9.3: Blind spot probability of the typical target node ($d = 25$ m, and $\lambda_0 = 0.015$ obstacles/km^2).

where $a = 1.07950, b = 3.03226$, and $c = 3.31122$. Using this result in (9.2), we obtain the asymptotic blind spot probability for the typical target. As expected, the blind spot probability computed using the analytical expressions matches exactly with the empirical results obtained from Monte Carlo simulations, as depicted in Fig. 9.3.

9.3   SUMMARY

In this chapter, we discussed the application of PLP to localization networks. The ability to localize a target depends on the visibility of the anchor nodes in the network. Modeling the obstacles in a network by a PLP, we characterized the distribution of the area of the visible region to the typical target node. Using this result, we computed the asymptotic blind spot probability of the typical target node.

CHAPTER 10

Path Distance Characteristics

In the previous chapters, we discussed the application of PLP and PLCP to wireless networks. In particular, we studied key performance metrics such as coverage probability and load distribution which are primarily characterized by the Euclidean distances between the points of the PLCP. In this chapter, we will mainly focus on the path distance characteristics of the PLCP. The analytical framework presented in this chapter can be leveraged to address several important questions in the areas of wireless networks, vehicular cyber-physical systems, transportation networks, and urban planning. While little is known about path distances in a PLCP, some explicit calculations are possible for a special case of PLCP called Manhattan Poisson line Cox process (MPLCP) which will be presented in this chapter. Specifically, we will derive the distribution of the distance to the nearest point of the MPLCP in the sense of path distance from the typical intersection of the MPLP, which will be defined shortly in Section 10.3. A brief introduction to MPLP and MPLCP are first provided in the following sections.

10.1 MANHATTAN POISSON LINE PROCESS

A MPLP is a variant of the PLP, where the orientation of the lines is restricted to $\{0, \frac{\pi}{2}, \pi, \frac{3\pi}{2}\}$, thereby forming a set of horizontal and vertical lines in $\mathbb{R}^2$. A MPLP Φ_l can be generated by four independent 1D PPPs Ψ_0, $\Psi_{\frac{\pi}{2}}$, Ψ_π, and $\Psi_{\frac{3\pi}{2}}$ along the lines $\theta = 0$, $\theta = \frac{\pi}{2}$, $\theta = \pi$, and $\theta = \frac{3\pi}{2}$, respectively, in the representation space $\mathcal{C}$. We denote the set of horizontal lines by $\Phi_{lh} \equiv \{L_{h_1}, L_{h_2}, \dots\}$ and the set of vertical lines by $\Phi_{lv} \equiv \{L_{v_1}, L_{v_2}, \dots\}$. A MPLP can also be interpreted as the set of lines obtained by drawing vertical and horizontal lines through the points populated on the x and y-axes according to independent 1D PPPs Ξ_x and Ξ_y, respectively. This interpretation will be useful in understanding some basic properties of MPLP, which will be discussed next.

10.1.1 STATIONARITY AND MOTION-INVARIANCE

From our discussion on planar line processes in Section 3.1.1, we already know that a line process is stationary if the distribution of lines is invariant to any translation $T_{(t,\beta)}$ which changes the coordinates of line in $\mathcal{C}$ from (ρ, θ) to $(\rho - t\cos(\theta - \beta), \theta)$. Therefore, a MPLP Φ_l is stationary if the corresponding 1D PPPs Ψ_0, $\Psi_{\frac{\pi}{2}}$, Ψ_π, and $\Psi_{\frac{3\pi}{2}}$ in $\mathcal{C}$ are stationary. Alternatively, Φ_l is stationary when Ξ_x and Ξ_y are stationary. As the orientation of the lines is not uniformly distributed in $[0, 2\pi)$, the MPLP is not isotropic and hence not motion-invariant.

10.1.2   LINES INTERSECTING A REGION

For a stationary MPLP Φ_l generated by independent and homogeneous 1D PPPs Ξ_x and Ξ_y with the same density λ_l, the number of vertical and horizontal lines that intersect a region $K \subset \mathbb{R}^2$ follows a Poisson distribution with mean $\lambda_l \nu_1(K_x)$ and $\lambda_l \nu_1(K_y)$, respectively, where K_x and K_y denote the projection of the region K onto x and y-axes, respectively.

10.2   MANHATTAN POISSON LINE COX PROCESS

Similar to PLCP, a MPLCP Φ_a is constructed by populating points on the lines of a MPLP Φ_l such that the locations of points on each line L form an independent 1D PPP Ψ_L. We limit our discussion to a MPLCP in which the underlying MPLP Φ_l is stationary and the 1D PPP Ψ_L is homogeneous with density λ_p. We will discuss some of the basic properties of the MPLCP next.

10.2.1   STATIONARITY AND MOTION-INVARIANCE

As we had discussed in Section 4.2, the distribution of the number of points of the MPLCP Φ_a that lie inside a set $A \subset \mathbb{R}^2$ depends on the distribution of the total length of line segments inside A, denoted by ℓ_A. Since ℓ_A is invariant to any translation of the plane due to the stationarity of Φ_l, the point process Φ_a is also stationary. Using a similar argument, we can also argue that Φ_a is not isotropic and hence not motion-invariant.

10.2.2   VOID PROBABILITY

The void probability of the MPLCP Φ_a is given in the following lemma.

Lemma 10.1   *The void probability of the MPLCP is*

$$\mathbb{P}(N_p(A) = 0) = \exp\left[-\lambda_l \int_\mathbb{R} 1 - \exp(-\lambda_p \nu_1(L_{\rho_h} \cap A)) \mathrm{d}\rho_h \right.$$
$$\left. -\lambda_l \int_\mathbb{R} 1 - \exp(-\lambda_p \nu_1(L_{\rho_v} \cap A)) \mathrm{d}\rho_v \right]. \quad (10.1)$$

Proof. The void probability can be computed as

$$\mathbb{P}(N_p(A) = 0) = \mathbb{E}_{\Phi_l}\left[\mathbb{P}(N_p(A) = 0 \mid \Phi_l) \right].$$

As the distribution of horizontal and vertical lines are independent of each other, the above equation can be written as

$$\mathbb{P}(N_p(A) = 0) = \mathbb{E}_{\Phi_{lh}}\left[\mathbb{P}(N_p(A) = 0 \mid \Phi_{lh}) \right] \mathbb{E}_{\Phi_{lv}}\left[\mathbb{P}(N_p(A) = 0 \mid \Phi_{lv}) \right]. \quad (10.2)$$

We will now consider the first term which corresponds to the distribution of points on the horizontal lines. As the distribution of points on the lines are independent, we have

$$\mathbb{E}_{\Phi_{lh}}\left[\mathbb{P}(N_p(A) = 0 \mid \Phi_{lh})\right] = \mathbb{E}_{\Phi_{lh}}\left[\prod_{\rho_h \in \Xi_y} \mathbb{P}(N_p(L_{\rho_h} \cap A) = 0)\right].$$

Using the expression for void probability of the 1D PPP and the PGFL of 1D PPP Ξ_y, the above equation simplifies to

$$\mathbb{E}_{\Phi_{lh}}\left[\mathbb{P}(N_p(A) = 0 \mid \Phi_{lh})\right] = \exp\left[-\lambda_l \int_{\mathbb{R}} 1 - \exp(-\lambda_p \nu_1(L_{\rho_h} \cap A)) d\rho_h\right].$$

Owing to the symmetry, we obtain a similar expression for the second term in (10.2) corresponding to the distribution of points on the vertical lines. Substituting these expressions in (10.2), we obtain the final expression for void probability.   □

10.3   THE SHORTEST PATH DISTANCE

Let us consider a stationary MPLCP Φ_a as described in Section 10.1. In this section, we will derive the distribution of the length of the shortest path from the typical intersection of the MPLP to its nearest point of the MPLCP in the sense of path distance, as depicted in Fig. 10.1. Owing to the stationarity of the MPLP, we place the origin at the location of the typical intersection without loss of generality. Thus, under the Palm distribution of the intersection points, the resulting line process is $\Phi_{l_0} \equiv \Phi_l \cup \{L_x\} \cup \{L_y\}$, where L_x and L_y denote the lines aligned along the x and y-axis, respectively. Consequently, under the Palm distribution, the resulting point process is $\Phi_{a_0} \equiv \Phi_a \cup \Psi_{L_x} \cup \Psi_{L_y}$.

Recall that the distribution of the nearest neighbor distance R for the PLCP was derived in Section 4.2.7 using counting measures, i.e., the probability that there are no points of the PLCP inside the disc $B(o, R)$. We will now adopt a similar approach in this case to derive the CDF of the shortest path distance. However, the key difference is that we are dealing with the path distance as opposed to the Euclidean distance in Section 4.2.7. Hence, we must carefully consider our region of interest in this case. For any point $x \in \Phi_a$ with coordinates (x_i, y_i) in $\mathbb{R}^2$, the length of its shortest path from the origin is given by $z_i = |x_i| + |y_i|$, which is the first order Minkowski distance of the point from the origin. Let us consider that the length of the shortest path to the nearest neighbor from the typical intersection at the origin is r_m. This means that there are no points in Φ_a whose path distance from the origin is smaller than r_m. In other words, there does not exist any point (x, y) in $\mathbb{R}^2$ such that $|x| + |y| < r_m$. Thus, the region of interest which cannot contain any point of the MPLCP is a square S_{r_m} formed by the intersection of the half-planes $x + y < r_m$, $-x + y < r_m$, $-x - y < r_m$, and $x - y < r_m$, as illustrated in Fig. 10.2. Using these conditions, the closed-form expression for the CDF of the

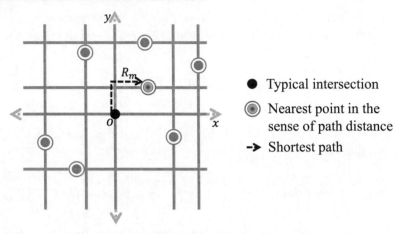

Figure 10.1: Illustration of the shortest path to the nearest point of the MPLCP in the sense of path distance from the typical intersection of the MPLP.

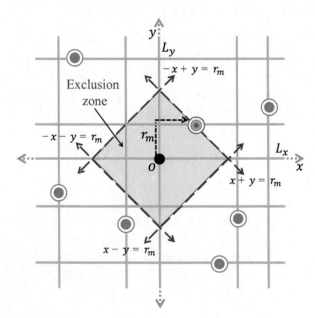

Figure 10.2: Illustration of the exclusion zone for the typical point at the intersection.

shortest path distance to the nearest point of the MPLCP in the path distance sense from the typical intersection is derived in the following theorem.

Theorem 10.2 The CDF of the shortest path distance from the typical intersection to the nearest point of the MPLCP

$$F_{R_m}(r_m) = 1 - \exp\left[-4\lambda_p r_m - 4\lambda_l r_m + \frac{2\lambda_l}{\lambda_p}\left(1 - e^{-2\lambda_p r_m}\right) \right]. \tag{10.3}$$

Proof. The CDF of R_m can be computed as

$$F_{R_m}(r_m) = 1 - \mathbb{P}(R_m > r_m) = 1 - \mathbb{P}(N_p(\Phi_{l_0} \cap S_{r_m}) = 0),$$

where $N_p(\cdot)$ denotes the number of points. We now need to compute the probability that there are no points on any of the lines intersecting the region S_{r_m}. In addition to the deterministic lines L_x and L_y, a random number of lines intersect S_{r_m}. As the point processes on different lines are independent, we will handle them separately. Thus, the above equation can be written as

$$F_{R_m}(r_m) = 1 - \mathbb{P}(N_p(L_x \cap S_{r_m}) = 0)\mathbb{P}(N_p(L_y \cap S_{r_m}) = 0)\mathbb{P}(N_p(\Phi_l \cap S_{r_m}) = 0). \tag{10.4}$$

From the void probability of the 1D PPP, the first two terms in the above equation are given by

$$\mathbb{P}(N_p(L_x \cap S_{r_m}) = 0) = \mathbb{P}(N_p(L_y \cap S_{r_m}) = 0) = \exp(-2\lambda_p r_m). \tag{10.5}$$

The third term in (10.4) can be computed using the expression for void probability given in Lemma 10.1, where we have to consider the region S_{r_m} in place of A. For a horizontal line at a distance $\rho_h < r_m$ from the origin, the length of the line segment inside the region S_{r_m} is $2(r_m - \rho_h)$. By symmetry, the length of a vertical line segment inside S_{r_m} at a distance ρ_v from the origin is $2(r_m - \rho_v)$. Using these properties in the expression for void probability, we obtain the third term in (10.4) as

$$\mathbb{P}(N_p(\Phi_l \cap S_{r_m}) = 0) = \exp\left[-\lambda_l \int_{-r_m}^{r_m} 1 - \exp(-2\lambda_p(r_m - \rho_h))\mathrm{d}\rho_h \right.$$
$$\left. -\lambda_l \int_{-r_m}^{r_m} 1 - \exp(-2\lambda_p(r_m - \rho_v))\mathrm{d}\rho_v \right]$$
$$= \exp\left[-4\lambda_l r_m + \frac{2\lambda_l}{\lambda_p}\left(1 - e^{-2\lambda_p r_m}\right) \right]. \tag{10.6}$$

Substituting (10.5) and (10.6) in (10.4), we obtain the final expression for the CDF of R_m. $\quad\square$

We computed the CDF of the length of the shortest path to the nearest point of the MPLCP from the typical intersection of the MPLP. Following a similar approach and additionally handling the coupling between multiple paths to a point, we can derive the CDF of the shortest path distance from the typical point of the MPLCP to its nearest neighbor in the path distance sense, which has recently been done in [63].

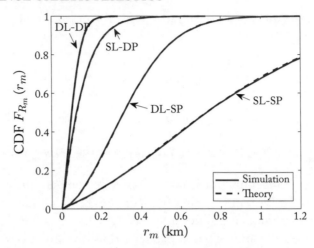

Figure 10.3: CDF of the path distance to the nearest neighbor of the MPLCP from the typical intersection of the MPLP: DL-DP ($\lambda_l = 15\,\text{km}^{-1}$, $\lambda_p = 2\,\text{km}^{-1}$), SL-DP ($\lambda_l = 2\,\text{km}^{-1}$, $\lambda_p = 2\,\text{km}^{-1}$), DL-SP ($\lambda_l = 15\,\text{km}^{-1}$, $\lambda_p = 0.1\,\text{km}^{-1}$), and SL-SP ($\lambda_l = 2\,\text{km}^{-1}$, $\lambda_p = 0.1\,\text{km}^{-1}$).

10.4   RESULTS AND DISCUSSION

In this section, we will first present some numerical results comparing the analytical results with the empirical CDF of the shortest path distance from the typical intersection to its nearest point of the MPLCP obtained through Monte Carlo simulations. We will then discuss the applications of these results to transportation networks, urban planning, and personnel deployment.

10.4.1   NUMERICAL RESULTS

We first simulate the spatial model described in Section 10.1 in MATLAB. For each realization of the model, we compute the shortest path distance to the points of the MPLCP from the typical intersection (at the origin) and then determine the minimum of those distances to obtain the shortest path distance to the nearest point of the MPLCP. This process is repeated for several realizations of the spatial model to compute the empirical CDF of the path distance to the nearest point of the MPLCP from the typical intersection. In order to visualize the path distance characteristics for various combinations of line and point densities, we evaluate these results under four broad regimes: (i) dense lines–dense points (DL-DP) corresponding to large values of λ_l and λ_p; (ii) dense lines–sparse points (DL-SP) corresponding to large values of λ_l and small values of λ_p; (iii) sparse lines–dense points (SL-DP) corresponding to small values of λ_l and large values of λ_p; and (iv) sparse lines–sparse points (SL-SP) corresponding to small values of λ_l and λ_p, as depicted in Fig. 10.3. As expected, the CDF obtained from Theorem 10.2 matches exactly with the empirical results in all four regimes.

10.4.2   APPLICATIONS

In addition to vehicular nodes and RSUs, other critical infrastructure such as gas stations or charging stations for electric vehicles that are located along the roads can be modeled by a PLCP or MPLCP. Therefore, in transportation networks, the analytical result presented in this chapter can be leveraged to obtain insights into the minimum travel distance between two places of interest. Furthermore, considering the typical intersection as the location of a vehicular user, we can obtain the distribution of the minimum travel distance of a user to the nearest facility of a certain type. We can also explore other metrics such as minimum travel duration or minimum fuel consumed to reach a desired destination. It is important to note that the minimum travel time to reach a facility also depends on the distribution of velocity of the vehicle on each road and hence, the nearest facility in the sense of path distance is not necessarily the one that can be reached with the least travel time. The insights gained from such analyses would be useful in infrastructure planning to ensure that the minimum travel time or minimum distance to a facility does not exceed a desired threshold. These results are also useful in the study of response time for first responders or police personnel to arrive at the site of an emergency. Hence, there is a lot of scope to further explore various metrics using this model.

10.5   SUMMARY

In this chapter, we derived the closed-form expression for the CDF of the path distance to the nearest point of the MPLCP (in the sense of the path distance) from the typical intersection of the MPLP. We also demonstrated the accuracy of analytical results numerically and discussed possible extensions of these results to study other metrics in various applications, such as transportation networks and urban planning. Modeling the locations of various facilities such as gas stations or charging stations for vehicles that are located along the roads by a MPLCP, we can obtain useful insights into the minimum travel distances and travel duration. The analytical framework presented in this chapter can be used to explore several such performance metrics. The approach presented in this chapter can also be extended to derive the CDF of the shortest path distance from the typical point of the MPLCP to its nearest neighbor in the path distance sense as done in [63].

CHAPTER 11

Potential Future Research

In this book, we concretely explained the theory of PLP and PLCP and discussed the analysis of basic performance metrics using these models in vehicular communication networks, localization networks, and transportation networks. As mentioned in Chapter 1, the research in the area of PLCP and its applications is still nascent and hence, several aspects of this model are yet to be explored. Building further on the knowledge base provided in this book, we now present key future directions of research from the perspective of spatial modeling, performance metrics, and mathematical analysis in vehicular communication networks. Naturally, this discussion is heavily influenced by the interests of the authors and is not meant to provide an exhaustive list of future directions of research.

11.1   ENHANCEMENTS OF THE MODEL

In this book, we focused entirely on the theory and applications for a motion-invariant PLCP. While this is a reasonable model that enables us to capture the spatial coupling between the vehicular nodes in a network, there is still a lot of scope for improving the model. Recall that the points of the PLCP on each line follow a homogeneous PPP. However, in reality, the density of vehicles on a road is often higher near the intersections than the remaining parts of the road. From the properties of the PLP, it is already known that the points of intersection of a line with other lines of the PLP form a 1D PPP [46, 47, 75]. Therefore, the concentration of vehicles at intersections on each line can be modeled as a 1D Poisson cluster process (PCP), where the cluster centers are located at the intersections. The performance analysis of vehicular communication networks using this model is a worthwhile extension of the current work.

   Another set of models that can be considered to improve the accuracy of spatial models for vehicular networks are hardcore processes. As the points of a PPP can be arbitrarily close to each other, it does not account for the physical separation between the vehicles due to the length of the vehicles and also the minimum safety distance maintained by the drivers. One of the solutions to address this is to simply model the distance between two neighboring vehicular nodes on each line of the PLP as the sum of an exponential random variable and a constant headway distance. The authors of [93, 94] have computed the mean and variance of interference measured at the typical vehicular user in this spatial model and have provided the approximate coverage probability. Several other aspects of this spatial model are yet to be investigated. One can also explore other spatial models, such as the Matérn hardcore point process, that can incorporate the constraints of minimum separation between the nodes.

11.2   PERFORMANCE METRICS

In this book, we mainly focused on the computation of SINR-based coverage probability for the typical receiver node in a vehicular communication network. The spatial model and the analysis presented in this book can be applied to study other key performance metrics such as rate coverage and latency. The rate coverage of the typical receiver is defined as the probability with which the data rate achievable at the typical receiver exceeds a predetermined threshold. This metric is particularly useful in vehicular communication networks because the vehicles may exchange media files of large sizes, such as the map of an area or a video footage of certain sections of the road. A high data throughput is often necessary to enable such services and hence, it is quite important to understand the rate coverage characteristics. In order to compute the rate coverage of the typical receiver, one of the key components is the distribution of the load on the serving node. In this book, we characterized the load on MBSs and RSUs due to vehicular users under the assumption that the RSUs serve only those vehicles that are located on its own road. Using these results, the rate coverage of the typical receiver in a C-V2X network has been recently computed in [37]. A meaningful extension of this work could be to analyze the rate coverage for a more general setup where the RSUs also serve the vehicular nodes located on the other roads. However, in order to accomplish this, we must first address some of the more fundamental open problems pertaining to the spatial model. For instance, since the load characteristics primarily depend on the statistical properties of the Voronoi cells formed by the PLCP, it would be quite useful to characterize some of the basic properties of the Poisson line Cox Voronoi tessellation, such as the area of the typical cell, which is still an open problem in the literature.

Another important metric that is critical to vehicular communication networks is the latency, i.e., the time taken from the generation of the message at the transmitter to the successful reception of the message at the receiver. The two key components of latency are the transmission delay, which is the wait time before the packet is transmitted from the antenna, and the local delay, which is caused by the retransmission of the packet due to packet failure. While the transmission delay can be characterized by using tools from queuing theory, the local delay primarily depends on the SINR characteristics and can be derived using the results presented in this book.

11.3   SPATIO-TEMPORAL ANALYSIS

In this book, we analyzed the performance of vehicular communication networks for a single snapshot of the network. However, due to the mobility of vehicles, the spatial configuration of the nodes in the network continuously changes with time and hence, it is of interest to analyze the temporal behavior of the network. Moreover, certain vehicular applications require a vehicle to maintain sustained connection to another vehicle or an RSU. For instance, let us consider the example of cooperative perception in fully autonomous vehicles where the goal is to extend a vehicle's perception beyond its field-of-view through the exchange of information with the nodes in its surrounding environment [95]. In this case, the vehicles need to remain connected to the other nodes in the network for a prolonged duration for safe maneuvering. Hence, in order to understand the performance of the network for such applications, it is necessary to include mobility of the vehicles in the analysis. For instance, we can model the in-flow of vehicles into a road as a Poisson arrival process and study the behavior of the network from one time instant to the next by considering the temporal correlation in the interference of the network. Using this setup, we can also study the handover rate of the moving vehicles between MBSs and RSUs. This is a promising direction for future work that can offer remarkable insights in the network design.

Bibliography

[1] D. J. Daley and D. Vere-Jones, *An Introduction to the Theory of Point Processes: Volume I Elementary Theory and Methods*, 2nd ed., New York, Springer, 2003. DOI: 10.1007/b97277. 1

[2] D. J. Daley and D. Vere-Jones, *An Introduction to the Theory of Point Processes: Volume II: General Theory and Structure*. Springer Science & Business Media, 2007. DOI: 10.1007/978-0-387-49835-5. 1, 2, 11

[3] R. Schneider and W. Weil, *Stochastic and Integral Geometry*. Berlin Heidelberg, Springer, 2008. DOI: 10.1007/978-3-540-78859-1. 1, 2

[4] S. N. Chiu, D. Stoyan, W. S. Kendall, and J. Mecke, *Stochastic Geometry and its Applications*, 3rd ed., John Wiley and Sons, 2013. DOI: 10.1002/9781118658222. 1, 2, 3, 7, 11, 21, 42

[5] J. Kingman, *Poisson Processes*. Oxford University Press, 1993. DOI: 10.1002/0470011815.b2a07042. 1

[6] M. Haenggi, *Stochastic Geometry for Wireless Networks*. Cambridge University Press, 2013. DOI: 10.1017/cbo9781139043816. 1, 7, 11, 14, 29, 33, 51, 80

[7] F. Baccelli and B. Błaszczyszyn, Stochastic geometry and wireless networks: Volume I theory, *NOW: Foundations and Trends in Networking*, 3(3–4):249–449, 2010. DOI: 10.1561/1300000006. 1, 11

[8] F. Baccelli and B. Błaszczyszyn, Stochastic geometry and wireless networks: Volume II applications, *NOW: Foundations and Trends in Networking*, 4(1–2):1–312, 2010. DOI: 10.1561/1300000026. 1

[9] M. Haenggi and R. K. Ganti, Interference in large wireless networks, *NOW: Foundations and Trends in Networking*, 3(2):127–248, February 2009. DOI: 10.1561/1300000015. 1

[10] R. Vaze, *Random Wireless Networks: An Information Theoretic Perspective*. Cambridge University Press, 2015. DOI: 10.1017/cbo9781316182581. 1

[11] S. Mukherjee, *Analytical Modeling of Heterogeneous Cellular Networks: Geometry, Coverage, and Capacity*. Cambridge University Press, 2014. DOI: 10.1017/cbo9781107279674. 1

[12] B. Błaszczyszyn, M. Haenggi, P. Keeler, and S. Mukherjee, *Stochastic Geometry Analysis of Cellular Networks*. Cambridge University Press, 2018. DOI: 10.1017/9781316667339. 1

[13] S. Weber and J. G. Andrews, Transmission capacity of wireless networks, *Foundations and Trends in Networking*, 5(2–3):109–281, 2012. DOI: 10.1561/1300000032. 1

[14] M. Haenggi, J. G. Andrews, F. Baccelli, O. Dousse, and M. Franceschetti, Stochastic geometry and random graphs for the analysis and design of wireless networks, *IEEE Journal on Selected Areas in Communications*, 27(7):1029–1046, 2009. DOI: 10.1109/jsac.2009.090902. 1, 59

[15] H. ElSawy, E. Hossain, and M. Haenggi, Stochastic geometry for modeling, analysis, and design of multi-tier and cognitive cellular wireless networks: A survey, *IEEE Communications Surveys and Tutorials*, 15(3):996–1019, 2013. DOI: 10.1109/surv.2013.052213.00000. 1, 59

[16] J. G. Andrews, A. K. Gupta, and H. S. Dhillon, A primer on cellular network analysis using stochastic geometry, *ArXiv Preprint*, 2016. arxiv.org/abs/1604.03183 1, 59

[17] H. ElSawy, A. Sultan-Salem, M. Alouini, and M. Z. Win, Modeling and analysis of cellular networks using stochastic geometry: A tutorial, *IEEE Communications Surveys and Tutorials*, 19(1):167–203, 2017. DOI: 10.1109/comst.2016.2624939. 1, 59

[18] M. Z. Win, P. C. Pinto, and L. A. Shepp, A mathematical theory of network interference and its applications, *Proc. of the IEEE*, 97(2):205–230, 2009. DOI: 10.1109/jproc.2008.2008764. 1, 59

[19] F. Baccelli and B. Błaszczyszyn, On a coverage process ranging from the boolean model to the Poisson–Voronoi tessellation with applications to wireless communications, *Advances in Applied Probability*, 33(2):293–323, 2001. DOI: 10.1239/aap/999188315. 1

[20] F. Baccelli, B. Błaszczyszyn, and P. Mühlethaler, Stochastic analysis of spatial and opportunistic ALOHA, *IEEE Journal on Selected Areas in Communications*, 27(7):1105–1119, 2009. DOI: 10.1109/jsac.2009.090908. 1

[21] J. G. Andrews, F. Baccelli, and R. K. Ganti, A tractable approach to coverage and rate in cellular networks, *IEEE Transactions on Communications*, 59(11):3122–3134, November 2011. DOI: 10.1109/tcomm.2011.100411.100541. 1

[22] H. S. Dhillon, R. K. Ganti, F. Baccelli, and J. G. Andrews, Modeling and analysis of K-tier downlink heterogeneous cellular networks, *IEEE Journal on Selected Areas in Communications*, 30(3):550–560, April 2012. DOI: 10.1109/jsac.2012.120405. 1

[23] S. Mukherjee, Distribution of downlink SINR in heterogeneous cellular networks, *IEEE Journal on Selected Areas in Communications*, 30(3):575–585, 2012. DOI: 10.1109/jsac.2012.120407. 1

[24] M. Di Renzo, Stochastic geometry modeling and analysis of multi-tier millimeter wave cellular networks, *IEEE Transactions on Wireless Communications*, 14(9):5038–5057, 2015. DOI: 10.1109/twc.2015.2431689. 1

[25] X. Lin, R. K. Ganti, P. J. Fleming and J. G. Andrews, Towards Understanding the Fundamentals of Mobility in Cellular Networks, in *IEEE Transactions on Wireless Communications*, 12(4):1686–1698, April 2013. DOI: 10.1109/twc.2013.022113.120506. 1

[26] P. Erdös and A. Rényi, On random graphs, I, *Publicationes Mathematicae (Debrecen)*, 6:290–297, 1959. 2

[27] P. Erdös and A. Rényi, On the evolution of random graphs, *Publication of the Mathematical Institute of the Hungarian Academy of Science,*, pages 17–61, 1960. 2

[28] D. J. Watts and S. H. Strogatz, Collective dynamics of "small-world" networks, *Nature*, 393(6684):440, 1998. DOI: 10.1515/9781400841356.301. 2

[29] R. Albert and A.-L. Barabási, Statistical mechanics of complex networks, *Reviews of Modern Physics*, 74:47–97, January 2002. DOI: 10.1103/revmodphys.74.47. 2

[30] D. J. Aldous and J. Shun, Connected spatial networks over random points and a route-length statistic, *Statistical Science*, 25(3):275–288, August 2010. DOI: 10.1214/10-sts335. 2

[31] C. Gloaguen, F. Fleischer, H. Schmidt, and V. Schmidt, Fitting of stochastic telecommunication network models via distance measures and Monte–Carlo tests, *Telecommunication Systems*, 31(4):353–377, April 2006. DOI: 10.1007/s11235-006-6723-3. 2, 4

[32] C. Gloaguen, F. Fleischer, H. Schmidt, and V. Schmidt, Analysis of shortest paths and subscriber line lengths in telecommunication access networks, *Networks and Spatial Economics*, 10(1):15–47, March 2010. DOI: 10.1007/s11067-007-9021-z. 2, 4

[33] F. Voss, C. Gloaguen, F. Fleischer, and V. Schmidt, Distributional properties of Euclidean distances in wireless networks involving road systems, *IEEE Journal on Selected Areas in Communications*, 27(7):1047–1055, September 2009. DOI: 10.1109/jsac.2009.090903. 2, 4

[34] C. Gloaguen, F. Fleischer, H. Schmidt, and V. Schmidt, Simulation of typical Cox Voronoi cells with a special regard to implementation tests, *Mathematical Methods of Operations Research*, 62(3):357–373, 2005. DOI: 10.1007/s00186-005-0036-2. 2, 4

[35] V. V. Chetlur and H. S. Dhillon, Coverage analysis of a vehicular network modeled as Cox process driven by Poisson line process, *IEEE Transactions on Wireless Communications*, 17(7):4401–4416, July 2018. DOI: 10.1109/twc.2018.2824832. 2, 4, 5, 47, 63, 65

[36] V. V. Chetlur, S. Guha, and H. S. Dhillon, Characterization of V2V coverage in a network of roads modeled as Poisson line process, *Proc., IEEE International Conference on Communications (ICC)*, pages 1–6, 2018. DOI: 10.1109/icc.2018.8422582. 2, 4, 47

[37] V. V. Chetlur and H. S. Dhillon, Coverage and rate analysis of downlink cellular vehicle-to-everything (C-V2X) communication, *IEEE Transactions on Wireless Communications*, 19(3):1738–1753, 2020. DOI: 10.1109/twc.2019.2957222. 2, 4, 5, 47, 81, 89, 101, 118

[38] V. V. Chetlur and H. S. Dhillon, Success probability and area spectral efficiency of a VANET modeled as a Cox process, *IEEE Wireless Communications Letters*, 7(5):856–859, October 2018. DOI: 10.1109/lwc.2018.2832199. 2, 4, 47

[39] V. V. Chetlur and H. S. Dhillon, Poisson line Cox process: Asymptotic characterization and performance analysis of vehicular networks, *Proc., IEEE Global Communications Conference (Globecom)*, 2019. DOI: 10.1109/globecom38437.2019.9014150. 2, 4, 5, 47

[40] C. Choi and F. Baccelli, An analytical framework for coverage in cellular networks leveraging vehicles, *IEEE Transactions on Communications*, 66(10):4950–4964, October 2018. DOI: 10.1109/tcomm.2018.2835456. 2, 4, 5, 47, 63, 65

[41] C. Choi and F. Baccelli, Poisson Cox point processes for vehicular networks, *IEEE Transactions on Vehicular Technology*, 67(10):10160–10165, October 2018. DOI: 10.1109/tvt.2018.2859909. 2, 4, 40, 47

[42] C. Choi and F. Baccelli, Spatial and temporal analysis of direct communications from static devices to mobile vehicles, *IEEE Transactions on Wireless Communications*, 18(11):5128–5140, 2019. DOI: 10.1109/twc.2019.2933393. 2, 4

[43] F. Baccelli, M. Klein, M. Lebourges, and S. Zuyev, Stochastic geometry and architecture of communication networks, *Telecommunication Systems*, 7(1):209–227, June 1997. DOI: 10.1023/A:1019172312328. 2, 4

[44] M. Kendall and P. Moran, *Geometrical Probability*, ser. Griffin's Statistical Monographs, Hafner Publishing Company, 1963. 3

[45] S. Goudsmit, Random distribution of lines in a plane, *Reviews of Modern Physics*, 17(2-3):321, 1945. DOI: 10.1103/revmodphys.17.321. 3

[46] R. E. Miles, Random polygons determined by random lines in a plane, *Proc. of the National Academy of Sciences*, 52(4):901–907, 1964. DOI: 10.1073/pnas.52.4.901. 3, 117

[47] R. E. Miles, Random polygons determined by random lines in a plane, II, *Proc. of the National Academy of Sciences*, 52(5):1157–1160, 1964. DOI: 10.1073/pnas.52.5.1157. 3, 117

[48] P. I. Richards, Averages for polygons formed by random lines, *Proc. of the National Academy of Sciences*, 52(5):1160–1164, 1964. DOI: 10.1073/pnas.52.5.1160. 3

[49] R. E. Miles, The various aggregates of random polygons determined by random lines in a plane, *Advances in Mathematics*, 10(2):256–290, 1973. DOI: 10.1016/0001-8708(73)90110-2. 3

[50] I. Crain and R. Miles, Monte–Carlo estimates of the distributions of the random polygons determined by random lines in a plane, *Journal of Statistical Computation and Simulation*, 4(4):293–325, 1976. DOI: 10.1080/00949657608810132. 3

[51] R. E. Miles, Poisson flats in Euclidean spaces Part I: A finite number of random uniform flats, *Advances in Applied Probability*, 1(2):211–237, 1969. DOI: 10.2307/1426218. 3

[52] R. E. Miles, Poisson flats in Euclidean spaces Part II: Homogeneous Poisson flats and the complementary theorem, *Advances in Applied Probability*, 3(1):1–43, 1971. DOI: 10.2307/1426328. 3

[53] M. Bartlett, The spectral analysis of point processes, *Journal of the Royal Statistical Society. Series B (Methodological)*, pages 264–296, 1963. DOI: 10.1111/j.2517-6161.1963.tb00508.x. 3

[54] M. Bartlett, The spectral analysis of two-dimensional point processes, *Biometrika*, 51(3/4):299–311, 1964. DOI: 10.1093/biomet/52.1-2.303-b. 3

[55] M. Bartlett, The spectral analysis of line processes, *Proc. 5th Berkeley Symposium on Mathematical Statistics and Probability*, 3:135–152, 1967. DOI: 10.1111/j.2517-6161.1963.tb00508.x. 3

[56] R. Davidson, Construction of line processes: Second order properties, *Izv. Akad. Nauk. Armjan. SSR Ser. Mat*, 5:219–234, 1970. 3

[57] F. Papangelou, Point processes on spaces of flats and other homogeneous spaces, *Mathematical Proceedings of the Cambridge Philosophical Society*, 80(2):297–314, Cambridge University Press, 1976. DOI: 10.1017/s0305004100052932. 3

[58] L. A. Santaló and I. Yanez, Averages for polygons formed by random lines in Euclidean and hyperbolic planes, *Journal of Applied Probability*, 9(1):140–157, 1972. DOI: 10.2307/3212643. 4

[59] A. Fairclough and G. Davies, Poisson line processes in 2 space to simulate the structure of porous media: Methods of generation, statistics, and applications, *Chemical Engineering Communications*, 92(1):23–48, 1990. DOI: 10.1080/00986449008911420. 4

[60] T. Meyer and H. H. Einstein, Geologic stochastic modeling and connectivity assessment of fracture systems in the boston area, *Rock Mechanics and Rock Engineering*, 35(1):23–44, February 2002. DOI: 10.1007/s006030200007. 4

[61] A. Rosenfeld and L. S. Davis, Image segmentation and image models, *Proc. of the IEEE*, 67(5):764–772, May 1979. DOI: 10.1109/proc.1979.11326. 4

[62] S. Aditya, H. S. Dhillon, A. F. Molisch, and H. Behairy, Asymptotic blind-spot analysis of localization networks under correlated blocking using a Poisson line process, *IEEE Wireless Communications Letters*, 6(5):654–657, October 2017. DOI: 10.1109/lwc.2017.2727490. 4, 107

[63] V. V. Chetlur, H. S. Dhillon, and C. P. Dettmann, Shortest path distance in Manhattan Poisson line Cox process, *ArXiv Preprint*, 2018. arxiv.org/abs/1811.11332 4, 113, 115

[64] F. Morlot, A population model based on a Poisson line tessellation, *Proc., Modeling and Optimization in Mobile, Ad Hoc and Wireless Networks*, pages 337–342, May 2012. 4, 40

[65] Y. Wang, K. Venugopal, A. F. Molisch, and R. W. Heath, MmWave vehicle-to-infrastructure communication: Analysis of urban microcellular networks, *IEEE Transactions on Vehicular Technology*, 67(8):7086–7100, August 2018. DOI: 10.1109/tvt.2018.2827259. 4

[66] C. Gloaguen, F. Voss, and V. Schmidt, Parametric distance distributions for fixed access network analysis and planning, *International Teletraffic Congress*, pages 1–8, September 2009. 4

[67] F. Baccelli and X. Zhang, A correlated shadowing model for urban wireless networks, *Proc., IEEE INFOCOM*, pages 801–809, 2015. DOI: 10.1109/infocom.2015.7218450. 4

[68] B. Błaszczyszyn and P. Mühlethaler, Random linear multihop relaying in a general field of interferers using spatial Aloha, *IEEE Transactions on Wireless Communications*, 14(7):3700–3714, July 2015. DOI: 10.1109/twc.2015.2409845. 4, 47

[69] G. Ghatak, A. De Domenico, and M. Coupechoux, Small cell deployment along roads: Coverage analysis and slice-aware RAT selection, *IEEE Transactions on Communications*, 67(8):5875–5891, August 2019. DOI: 10.1109/tcomm.2019.2916794. 4, 47

[70] S. Singh, H. S. Dhillon, and J. G. Andrews, Offloading in heterogeneous networks: Modeling, analysis, and design insights, *IEEE Transactions on Wireless Communications*, 12(5):2484–2497, May 2013. DOI: 10.1109/twc.2013.040413.121174. 14, 89

[71] E. N. Gilbert, Random subdivisions of space into crystals, *Annals of Mathematical Statistics*, 33(3):958–972, September 1962. DOI: 10.1214/aoms/1177704464. 14, 15, 19

[72] J. Meijering, Interface area, edge length, and number of vertices in crystal aggregates with random nucleation, *Philips Research Reports*, 8:270–290, 1953. 15, 16

[73] S. Corrsin, A measure of the area of a homogeneous random surface in space, *Quarterly of Applied Mathematics*, 12(4):404–408, 1955. DOI: 10.1090/qam/65947. 15, 19

[74] C. S. Smith and L. Guttman, Measurement of internal boundaries in three-dimensional structures by random sectioning, *JOM*, 5(1):81–87, 1953. DOI: 10.1007/bf03397456. 19

[75] L. Santaló, *Introduction to Integral Geometry*, ser. Actualités Scientifiques et Industrielles, Hermann, 1953. DOI: 10.1002/zamm.19790590633. 19, 29, 117

[76] R. R. A. Morton, The expected number and angle of intersections between random curves in a plane, *Journal of Applied Probability*, 3(2):559–562, 1966. DOI: 10.2307/3212140. 19

[77] L. Muche and D. Stoyan, Contact and chord length distributions of the Poisson Voronoi tessellation, *Journal of Applied Probability*, 29(2):467–471, 1992. DOI: 10.2307/3214584. 21

[78] M. Tanemura, Statistical distributions of Poisson Voronoi cells in two and three dimensions, *Forma*, pages 221–247, 2003. 37, 98, 107

[79] H. Hartenstein and L. P. Laberteaux, A tutorial survey on vehicular ad hoc networks, *IEEE Communications Magazine*, 46(6):164–171, June 2008. DOI: 10.1109/mcom.2008.4539481. 47

[80] S. Biswas, R. Tatchikou, and F. Dion, Vehicle-to-vehicle wireless communication protocols for enhancing highway traffic safety, *IEEE Communications Magazine*, 44(1):74–82, January 2006. DOI: 10.1109/mcom.2006.1580935. 47

[81] P. Papadimitratos, A. D. L. Fortelle, K. Evenssen, R. Brignolo, and S. Cosenza, Vehicular communication systems: Enabling technologies, applications, and future outlook on intelligent transportation, *IEEE Communications Magazine*, 47(11):84–95, November 2009. DOI: 10.1109/mcom.2009.5307471. 47

[82] 3GPP TR 36.885, Study on LTE-based V2X services, July 2016. 47

[83] 3GPP TR 38.885, Study on architecture enhancements for 3GPP support of advanced V2X services, March 2019. 47

[84] H. S. Dhillon and J. G. Andrews, Downlink rate distribution in heterogeneous cellular networks under generalized cell selection, *IEEE Wireless Communications Letters*, 3(1):42–45, February 2014. DOI: 10.1109/wcl.2013.110713.130709. 80

[85] J. P. Jeyaraj and M. Haenggi, A transdimensional Poisson model for vehicular networks, *Proc., IEEE Global Communications Conference (Globecom)*, 2019. DOI: 10.1109/globecom38437.2019.9014157. 86

[86] J. G. Andrews, S. Singh, Q. Ye, X. Lin, and H. S. Dhillon, An overview of load balancing in hetnets: Old myths and open problems, *IEEE Wireless Communications*, 21(2):18–25, 2014. DOI: 10.1109/mwc.2014.6812287. 89

[87] G. P. Patil, Weighted distributions, *Wiley StatsRef: Statistics Reference Online*, 2014. DOI: 10.1002/9781118445112.stat07359. 91, 99, 100

[88] P. D. Mankar, P. Parida, H. S. Dhillon, and M. Haenggi, Distance from the nucleus to a uniformly random point in the typical and the Crofton cells of the Poisson-Voronoi tessellation, 2019, available online arxiv/abs/1907.03635 97

[89] A. Chattopadhyay, B. Błaszczyszyn, and E. Altman, Two-tier cellular networks for throughput maximization of static and mobile users, *IEEE Transactions on Wireless Communications*, 18(2):997–1010, February 2019. DOI: 10.1109/twc.2018.2887386. 101

[90] T. Bai, R. Vaze, and R. W. Heath, Using random shape theory to model blockage in random cellular networks, *International Conference on Signal Processing and Communications (SPCOM)*, pages 1–5, July 2012. DOI: 10.1109/spcom.2012.6290250. 105

[91] T. Bai, R. Vaze, and R. W. Heath, Analysis of blockage effects on urban cellular networks, *IEEE Transactions on Wireless Communications*, 13(9):5070–5083, September 2014. DOI: 10.1109/twc.2014.2331971. 105

[92] A. Hinde and R. Miles, Monte Carlo estimates of the distributions of the random polygons of the Voronoi tessellation with respect to a Poisson process, *Journal of Statistical Computation and Simulation*, 10(3-4):205–223, 1980. DOI: 10.1080/00949658008810370. 107

[93] K. Koufos and C. P. Dettmann, Moments of interference in vehicular networks with hardcore headway distance, *IEEE Transactions on Wireless Communications*, 17(12):8330–8341, December 2018. DOI: 10.1109/twc.2018.2876241. 117

[94] K. Koufos and C. P. Dettmann, Performance of a link in a field of vehicular interferers with hardcore headway distance, *ArXiv Preprint*, 2018. arxiv.org/abs/1810.00959 117

[95] S. Kim, B. Qin, Z. J. Chong, X. Shen, W. Liu, M. H. Ang, E. Frazzoli, and D. Rus, Multivehicle cooperative driving using cooperative perception: Design and experimental validation, *IEEE Transactions on Intelligent Transportation Systems*, 16(2):663–680, April 2015. DOI: 10.1109/tits.2014.2337316. 119

[96] M. N. Sial, Y. Deng, J. Ahmed, A. Nallanathan, and M. Dohler, Stochastic geometry modeling of cellular V2X communication over shared channels, in *IEEE Transactions on Vehicular Technology*, 68(12):11873–11887, December 2019. DOI: 10.1109/TVT.2019.2945481. 47

Authors' Biographies

HARPREET S. DHILLON

Harpreet S. Dhillon received the B.Tech. degree in electronics and communication engineering from IIT Guwahati in 2008, the M.S. degree in electrical engineering from Virginia Tech in 2010, and the Ph.D. degree in electrical engineering from the University of Texas at Austin in 2013. After serving as a Viterbi Postdoctoral Fellow at the University of Southern California for a year, he joined Virginia Tech in 2014, where he is currently an Associate Professor of electrical and computer engineering and the Elizabeth and James E. Turner Jr. '56 Faculty Fellow. His research interests include communication theory, wireless networks, stochastic geometry, and machine learning. He is a Clarivate Analytics Highly Cited Researcher and has coauthored five best paper award recipients including the 2014 IEEE Leonard G. Abraham Prize, the 2015 IEEE ComSoc Young Author Best Paper Award, and the 2016 IEEE Heinrich Hertz Award. He was named the 2017 Outstanding New Assistant Professor, the 2018 Steven O. Lane Junior Faculty Fellow, the 2018 College of Engineering Faculty Fellow, and the recipient of the 2020 Dean's Award for Excellence in Research by Virginia Tech. His other academic honors include the 2008 Agilent Engineering and Technology Award, the UT Austin MCD Fellowship, and the 2013 UT Austin Wireless Networking and Communications Group leadership award. He currently serves as an Editor or Senior Editor for three IEEE journals.

VISHNU VARDHAN CHETLUR

Vishnu Vardhan Chetlur received the B. E. (Hons.) degree in Electronics and Communications Engineering from the Birla Institute of Technology and Science (BITS) Pilani, India, in 2013. After his graduation, he worked as a design engineer at Redpine Signals Inc. for two years. He is currently a Ph.D. student at Virginia Tech, where his research interests include wireless communication, stochastic geometry, vehicular networks, and smart cities. He graduated top of his class in the Department of Electrical Engineering at BITS and was awarded the Silver medal for being ranked second in the whole institute. He was also a recipient of the Pratt Scholarship at Virginia Tech and BITS merit scholarship for his excellence in academics. He has held internship positions at Philips Research India in Bangalore, India, Qualcomm Technologies Inc. in San Diego, CA, and Nokia Bell Labs in Murray Hill, NJ.

Printed in the United States
by Baker & Taylor Publisher Services